Studies in Systems, Decision and Control

Volume 447

Series Editor

Janusz Kacprzyk, Systems Research Institute, Polish Academy of Sciences,
Warsaw, Poland

The series "Studies in Systems, Decision and Control" (SSDC) covers both new developments and advances, as well as the state of the art, in the various areas of broadly perceived systems, decision making and control–quickly, up to date and with a high quality. The intent is to cover the theory, applications, and perspectives on the state of the art and future developments relevant to systems, decision making, control, complex processes and related areas, as embedded in the fields of engineering, computer science, physics, economics, social and life sciences, as well as the paradigms and methodologies behind them. The series contains monographs, textbooks, lecture notes and edited volumes in systems, decision making and control spanning the areas of Cyber-Physical Systems, Autonomous Systems, Sensor Networks, Control Systems, Energy Systems, Automotive Systems, Biological Systems, Vehicular Networking and Connected Vehicles, Aerospace Systems, Automation, Manufacturing, Smart Grids, Nonlinear Systems, Power Systems, Robotics, Social Systems, Economic Systems and other. Of particular value to both the contributors and the readership are the short publication timeframe and the world-wide distribution and exposure which enable both a wide and rapid dissemination of research output.

Indexed by SCOPUS, DBLP, WTI Frankfurt eG, zbMATH, SCImago.

All books published in the series are submitted for consideration in Web of Science.

Hsu-Shih Shih · David L. Olson

TOPSIS and its Extensions: A Distance-Based MCDM Approach

Springer

Hsu-Shih Shih
Department of Management Sciences
Tamkang University
New Taipei, Taiwan

David L. Olson
Department of Supply Chain Management
and Analytics, College of Business
University of Nebraska–Lincoln
Lincoln, NE, USA

ISSN 2198-4182 ISSN 2198-4190 (electronic)
Studies in Systems, Decision and Control
ISBN 978-3-031-09576-4 ISBN 978-3-031-09577-1 (eBook)
https://doi.org/10.1007/978-3-031-09577-1

This Springer imprint is published by the registered company Springer Nature Switzerland AG
The registered company address is: Gewerbestrasse 11, 6330 Cham, Switzerland

Preface

In the real world, there are many problems that need to be dealt with and solved with a satisfactory solution. This book makes an extensive investigation of the Technique for Order Preference by Similarity to Ideal Solution (TOPSIS), a prominent distance-based Multi-Criteria Decision-Making (MCDM) method, and its variations and possible extensions.

MCDM is important as it provides tools for decision-makers to assess tradeoffs in alternative performance. TOPSIS is widely used, probably due to its reliance on the data rather than exhaustive analysis of decision-makers' preference. A number of methods are thoroughly reviewed in the book. Also presented are many real-world examples taken from the published literature. The book intends to give a thorough and rigorous analysis of methods along with a demonstration of practical examples.

Objective of the Study

The objective herein aims to provide sufficient materials to demonstrate the development of TOPSIS up to date and to be used as a handbook on it. It contains the basic process of TOPSIS, numerous variant processes, property explanations, theoretical developments, and illustrative examples with real-world cases. Possible readers would be graduate students, researchers, analysts, and professionals, who are interested in TOPSIS, a distance-based algorithm, or who would like to compare TOPSIS with other MCDM methods when dealing with decision-making on finite numbers of alternatives. Aside from deliberation for research reference, its arrangement could be employed as a self-learning book with step-by-step illustrations for the MCDM community. Due to space constraints and personal concentration, coverage is limited to the crisp domain with traditional preference structure, in which fuzzy and other domains are mentioned only in simple forms. Even so, we believe the materials cover the essential issues in this area.

Book Organization

TOPSIS has proven highly popular, with a strong research stream that continues to grow each year. This research stream has seen many variants of TOPSIS. This book gives a theoretical review of the basic TOPSIS method, explaining each step of the TOPSIS process supported by the presentation of computations for real examples taken from the published literature. Chapter 1 discusses the process of multiple criteria modeling. Chapter 2 describes how TOPSIS fits into that process and presents the TOPSIS algorithm and its relationship to decision-maker choice behavior. Chapter 3 presents basic variants of the TOPSIS method. TOPSIS support for group decision-making is discussed in Chap. 4. Chapter 5 discusses decision support system aspects of TOPSIS. TOPSIS can be used for multiple tasks. Chapter 6 focuses on the use of TOPSIS for sorting problems, as opposed to identification of one preferred alternative. Chapter 7 deals with the potential for rank reversal in TOPSIS. We noted that the stream of TOPSIS research has led to the presentation of many variants. Chapter 8 discusses methods closely related to the TOPSIS idea of getting close to the ideal and far away from the nadir. Fuzzy TOPSIS methods model decisions from the perspective of imprecise measures. VIKOR applied the TOPSIS idea using Minkowski metrics. The Best-Worst Method (BMW) focuses on using the best and worst to generate stable criteria weights. The book concludes with Chap. 9, which reviews TOPSIS applications published in the literature.

Hwang CL, Yoon K (1981). Multiple Attribute Decision Making: A State-of-the-Art Survey. Springer-Verlag, Berlin. ISBN: 9783540105589.

New Taipei, Taiwan Hsu-Shih Shih
Lincoln, USA David L. Olson

Acknowledgements

The first author has known TOPSIS ever since he studied at Kansas State University, U.S.A., but found many ambiguous works and associated extensions after teaching MCDM courses in Taiwan, R.O.C. He thus has attempted to bridge the gap between existing works and the essence of TOPSIS. Over the past two decades, many colleagues and students have inspired him to investigate the insights of TOPSIS. He is thankful to Prof. Huan-Jyh Shyur at Tamkang University, Taiwan, for his enduring cooperation and support. He is also grateful to Prof. David L. Olson of the University of Nebraska-Lincoln, the second author, for his long-term support and for acting as a mentor to clarify his thoughts. With the help of Profs. Ignacy Kaliszewski and Dmitry Podkopaev at the Polish Academy of Sciences, he has been able to involve the properties of TOPSIS. He also appreciates those pioneers in shaping TOPSIS, including Prof. Ching-Lai Hwang at Kansas State University, the founder of the method, Prof. Po Lung Yu at National Chiao Tung University and the University of Kansas, and Prof. Milan Zeleny at Fordham University, U.S.A.

During his own development spanning two decades, numerous professors and friends have supported him in preparing the contents, including Profs. E. Stanley Lee, Shing I. Chang, and David Ben-Arieh at Kansas State University, Dr. Young-Jou Lai at Halliburton, Texas, Prof. Minghe Sun at the University of Texas at San Antonio, Prof. Jennifer S. Shang at the University of Pittsburgh, U.S.A., Prof. Luiz Flavio Autran Monteiro Gomes at Ibmec, Brazil, Prof. Przemysław Juszczuk at the University of Economics in Katowice, Poland, Prof. Bertrand Mareschal at Université Libre de Bruxelles, Belgium, Prof. Gwo-Hshiung Tzeng at National Taipei University, Profs. Hsiao-Fan Wang and Ue-Pyng Wen at National Tsing Hua University, Prof. Chiang Kao at National Cheng Kung University, Prof. Liang-Chih Huang at National Chung Cheng University, Prof. Ping-Teng Chang at Tunghai University, Prof. Ping-Feng Pai at National Chi Nan University, Prof. Jei-Zheng Wu at Soochow University, Prof. Chih-Hung Wang at National Chiayi University, and Profs. Hai-Ming Chen, I-Fei Chen, Chi-Bin Cheng, Wen-Shuenn Deng, Chao-Che Hsu, Chichang Jou, Tong-liang Kao, Jyh-Jiuan Lin, Hui-Chiung Lo, Jerry C.Y. Miao, Liang-Yuh Ouyang, Tzong-Ru Tsai, Ruey-Chyn Tsaur, Wei-Tzen Yang, and Dr. Yuan-Sheng Lee at Tamkang University, Taiwan. Various students have also contributed to this book, especially Mrs. Shang-Yu Chung, Wen-Yuan Lin, Hung-Ching Hu, Chong-Ming Yan, and Miss Min-Chen Kan.

He is grateful to the Ministry of Science and Technology, Taiwan, R.O.C., for the financial support for two decades and the Ministry of National Defense for funding his Ph.D. study in the U.S.A. and for offering a chance for him to dive into the area of MCDM. Special thanks are given to Tamkang University and I-Shou University, Taiwan, and three foreign visiting institutions: the University of Pittsburgh (hosted by Prof. Thomas L. Saaty), Aalto University (conducted by Prof. Raimo P. Hämäläinen), and the Norwegian School of Economics (hosted by Prof. Stein W. Wallace), for supporting ideas with inspiration, research facilities, and general assistance.

New Taipei, Taiwan, R.O.C. — Hsu-Shih Shih
Lincoln, Nebraska, USA — David L. Olson
July 2022 — James & H.K. Stuart Chancellor's Distinguished Chair

Contents

About the Authors

Hsu-Shih Shih is the Professor at the Department of Management Sciences, Tamkang University, Taiwan, R.O.C. He obtained his Ph.D. degree at the Department of Industrial and Manufacturing Systems Engineering of Kansas State University, Manhattan, Kansas, U.S.A., in 1995, MS degree at the Department of Mechanical Engineering at Chung Cheng Institute of Technology, Taiwan in 1984, and BS degree at the Department of Naval Architecture and Marine Engineering of the same institute in 1978. He has also been a visiting scholar at the University of Pittsburgh, U.S.A. (2007), Aalto University, Finland (2010), and the Norwegian School of Economics, Norway (2014). His areas of specialization include decision analysis and support, operations research, soft computing, and environmental management. His publications appear in the European Journal of Operational Research, Computers and Operations Research, Journal of Operational Research Society, Central European Journal of Operations Research, Operational Research: An International Journal, Group Decision and Negotiation, Fuzzy Sets and Systems, Journal of Cleaner Production, Information Sciences, Computers and Mathematics with Applications, Mathematical and Computer Modelling, Applied Mathematics Letters, etc., along with four book chapters, and three books by Springer. He has been Secretary (2012–2013), Council Member (2014–2015), President (2016–2017), and Supervisor (2018–2019) of the Operations Research Society of Taiwan, as well as a member of the executive committee of International Society on Multiple Criteria Decision Making (2017–2021).

David L. Olson is the James & H.K. Stuart Professor and Chancellor's Professor at the University of Nebraska. He has published research in over 200 refereed journal articles, primarily on the topic of multiple objective decision-making, information technology, supply chain risk management, and data mining. He teaches in management information systems, management science, and operations management areas. He has authored over 40 books, to include Decision Aids for Selection Problems, Introduction to Information Systems Project Management, Managerial Issues of Enterprise Resource Planning Systems, Supply Chain Risk Management, and Supply

Chain Information Technology. Additionally, he has co-authored the books Introduction to Business Data Mining, Enterprise Risk Management, Advanced Data Mining Techniques, Enterprise Information Systems, Enterprise Risk Management Models, and Financial Enterprise Risk Management. He has served as associate editor of Service Business, Decision Support Systems, and Decision Sciences and co-editor in chief of the International Journal of Services Sciences. He has made over 200 presentations at international and national conferences on research topics. He is a member of the Decision Sciences Institute, the Institute for Operations Research and Management Sciences, and the Multiple Criteria Decision Making Society. He was a Lowry Mays endowed Professor at Texas A&M University from 1999 to 2001. He was named the Raymond E. Miles Distinguished Scholar award for 2002 and was a James C. and Rhonda Seacrest Fellow from 2005 to 2006. He was named Best Enterprise Information Systems Educator by IFIP in 2006. He is a Fellow of the Decision Sciences Institute.

List of Figures

Chapter 1
Multiple Criteria Modeling

The real world has many problems to be solved, including selection decisions, and one initial step is to identify problem characteristics and obtain their measures. This is termed the modeling process. Such decisions often involve multiple criteria. This chapter addresses the issues related to modeling through such multiple criteria.

1.1 Problems

A problem is a question raised for inquiry, consideration, or solution and generally occurs when humans feel there is a difference or discrepancy between the existing situation and the situation desired. People may feel uncomfortable if the problem remains unsolved or an undesired situation exists. For instance, a quality manager could set a yield rate for the manufacturing of integrated circuit wafers at 99.9999%, but the actual rate is below that standard. A financial manager could target the rate of return (ROR) for his or her company next year at 20%, yet there could be a problem if the actual ROR obtained were to be 10%. Of course, if ROR hits a higher standard, then this is a separate problem, but the firm is happy to accept it. There is still a need to examine this out-performance problem and better understand how it arose.

We sometimes have trouble identifying and treating the wrong problem. It is always important to carefully recognize the real problem, which could be difficult to see Clemen (1996). Aside from knowing the environment well, we might need help from experienced experts with domain knowledge. Some tools, such as 5W1H (What? Who? Where? When? Why? and How?) (Robertson 1946), brain storming (Osborn 1953), Ishikawa diagram or fishbone diagram (Ishikawa 1968), TRIZ (a Russian acronym for Theory of the Resolution of Invention-related Tasks) (Altshuller and Shapiro 1956), or other creative techniques, could also help to find the real problem. Aside from quantitative data illustrating the problem in a clear

H.-S. Shih and D. Olson, *TOPSIS and its Extensions: A Distance-Based MCDM Approach*, Studies in Systems, Decision and Control 447,
https://doi.org/10.1007/978-3-031-09577-1_1

style, qualitative data or even a figure could offer an image of the possible problems that exist. Some graphic tools such as the fishbone diagram help us to catch the problem, not in an exact way, but roughly capturing the system components of the problem. Such methods are helpful, especially for the early stages of decision making. In addition, a hierarchy or network could assist in visualizing the problem and its relevant elements through a geometric representation (Saaty and Shih 2009).

A problem can be classified based on the degree of the structured evidence (Turban et al. 2006). This classification falls along a continuum that ranges from highly structured to highly unstructured. The structured problem is routine and typically repetitive, whereas the unstructured problem is novel or uncommon, with ambiguous or insufficient information that makes it difficult to locate and solve. Although there is no clear boundary in the classification, semi-structured problems in the middle of the continuum can contain partial characteristics of the two extremes.

Each problem has an owner, who is the one affected by the issues to be solved or who indicates those who would benefit from the solution. Stakeholder is a more general term in decision making that includes problem owner(s) or decision makers (DMs), problem user(s) or solution executor(s), problem customer(s) or consequence beneficiaries or victim(s), and problem solver(s) or analyst(s) (Daellenbach 1994). All can be involved and are important for any further process. It is noteworthy that when the value of the problem or the stake is low, just like much in our daily lives, we usually deal with it by intuition; otherwise, we may process it through a formal procedure, generally guided by rationality.

1.2 Problem-Solving Process

After finding a problem, or a stimulus, the DM would provide a response and try to solve it, and the response initiates a problem-solving process. Simon (1977) provided an idea on the decision-making process, related to the problem-solving process, with the phases of: (i) intelligence, (ii) design, and (iii) choice. He later added another (iv) implementation (Turban et al. 2006). The intelligence phase is for scanning the environment of the problem and the opportunities; the second phase involves invention, development, and analysis of possible courses of actions; the third phase implies a selection from possible actions; and the fourth phase is associated with putting a solution into action. The process covers identifying the problem to providing a solution for the problem. From the perspective of management, the process usually involves a monitoring phase so that we can examine the consequences of implementation and confirmation that the consequences were as expected; otherwise, we need another cycle of the process for problem solving.

Some experts consider the cognitive role of the DM. Rowe and Boulgarides (1992) added a stimulus phase to proceed with the process given above. The cognitive phase considers external forces as opportunities, threats, or feedback of the existing system, which are the source of the problem. In marketing, they defined different phases of moment of truth (MOT) for on-line purchasing behavior. Less than zero MOT (<ZMOT) involves a stimulus to consumers, which is the awareness of need (Hyken 2016). Later, zero MOT or ZMOT indicates that the prospect recognizes this need and gathers information for potential purchase (Google 2012). The purchase decision happens at the first MOT or FMOT. There are another two phases related to post-sale experience and sharing experiences. These views detail the mental and behavior actions involved during the decision-making process. In addition to the DM's cognition, decision theory focuses on preference, which is a technical word for choosing between alternatives (Wikipedia). It closely relates to desires and involves a comparison of two alternatives in order to make a decision.

It would be interesting to differentiate the terms of problem solving and decision making. Some consider the first three phases are the decision-making process and the entire four phases the problem-solving process (Turban et al. 2006). In other words, the decision-making process ends with a choice and recommendation of a preferred solution, and the problem-solving process includes the execution of the solution.

Experts of operations research have extended (Simon's 1977) concept in a similar manner. The principal phases of an operations research study include: (i) definition of the problem, (ii) construction of the model, (iii) solution of the model, (iv) validation of the model, and (v) implementation of the solution (Taha 2017). The contents in the definition of the problem are usually vague, and the approach heavily relies on analysts' or DMs' perceptions and experiences. Thus, the approach is more art than science. Furthermore, especially for complicated challenges, the decision-making process is usually carried out by a group. Some common techniques, such as value-focused thinking, could facilitate common understanding of a problem situation and help DMs better understand the alternatives and their priorities, justify them, legitimize their findings, and facilitate the validation process (Bouyssou et al. 2006). We illustrate these steps in Fig. 1.1, where validation and verification have different purposes.

1.3 Multiple Criteria Formulation

When faced with a problem, a rational systematic approach is to formulate the problem mathematically. The formulation aims to identify the essence of the problem with bounded or limited rationality (Simon 1955). This often involves multiple criteria to capture the characteristics (factors, properties, value dimensions, or attributes) of the problem through multiple aspects. For example, jobs may be characterized by prestige, location, salary, and advancement opportunities (MacCrimmon 1968). If we think further, then criteria are usually in conflict, making the decision difficult. Cost or price is customarily an important criterion; the measure of quality

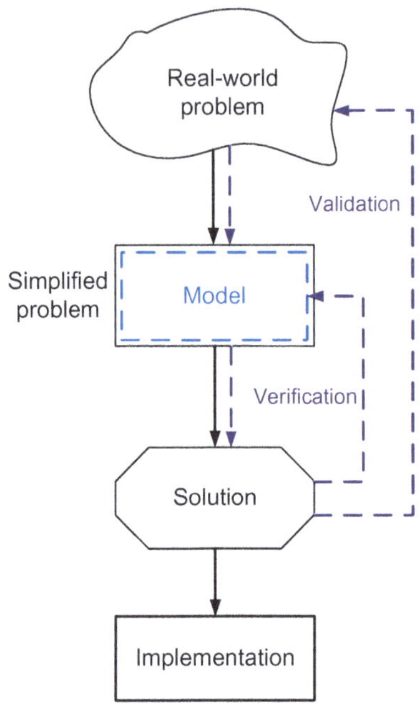

Fig. 1.1 Problem-solving process

is typically another factor that is easily in conflict with cost. When purchasing a car, cost, comfort, safety, and fuel economy are usually important, and it is rare that the cheapest car is also the most comfortable or the safest. It is noted that the multiple criteria formulation in this book focuses on traditional preference structure (Vincke 1992) and ignores fuzzy and other preference structures.

Such criteria could be first identified through common understanding, and some statistical tools could be helpful to identify the key criteria or elements in the formulation. For instance, principal component analysis (PCA) is often used to reduce the dimensionality of large datasets by transforming a large set of variables into a smaller set that still contains most of the information in the large set while ignoring the rest (Jolliffe 2002). Keeney and Raiffa (1993) suggested that the desirable properties of a set of criteria or attributes are: completeness, operational, decomposable, non-redundancy, and minimum size. These five properties offer a guideline for the selection of the necessary criteria. Wang et al. (2009) reviewed criteria used in a sustainable energy application and gave five principles to select major or important criteria: systemic, consistency, independency, measurability, and comparability. In general, the important criteria selected should cover the characteristics of the problem. After the analysis, other criteria could be left for further processing. To make the formulation process easier, independency on criteria is assumed in most multiple criteria formulations. If the phenomenon is noteworthy, however, the dependence of the criteria should be taken into account. We shall go into more detail on this topic in later chapters.

Criteria have their own essential characteristics. Some imply that the more, the better, such as high quality or reliability being more preferable than vice versa. These are classified as benefit criteria, and their utility or value increases monotonically. With other criteria, more is worse, such as oil consumption being less preferable to no oil consumption. They are classified as cost criteria and deliver decreasing monotonic utility. Yoon and Hwang (1995) classified these criteria into three groups: benefit, cost, and non-monotonic. Non-monotonic criteria offer non-monotonic utility with a fixed value that is the most preferable, and divergence from this value, more or less, is less preferred. A bell-shaped function is one kind of non-monotonic function. Examples include room temperature in an office or blood sugar level in a human body (Hwang and Yoon 1981).

The multi-criteria problem defines a set of alternatives $\boldsymbol{A} = \{A_1, A_2, \ldots, A_m\}$ with a set of evaluation criteria $\boldsymbol{F} = \{X_1, X_2, \ldots, X_n\}$ on $\boldsymbol{A}$, in which a DM wishes to solve one of the following three types of problems (Vincke 1992).

(i) Determine a subset of $\boldsymbol{A}$ considered to be the best with respect to $\boldsymbol{F}$ (choice problem);
(ii) Rank the elements of $\boldsymbol{A}$ from the best to the worst (ranking problem).
(iii) Divide $\boldsymbol{A}$ into subsets according to some norms (sorting problem).

Though the first two types of problems are only slightly different, solutions to either one could be obtainable on condition that we can rank all alternatives in an order from best to worst. The selection of the best one within the order is the choice problem. Since the founded norms are diverse in a multi-criteria problem, sorting could generate a problem with quite different results under different criteria. This type has a close relation with classification and clustering in data analysis (Jajuga et al. 2002), and the number of subsets could be known or unknown in advance. The resulting subsets should be ordered by any ranking index. Moreover, the evaluation is generally given to a decision table or matrix, evaluation matrix, or project impact matrix, in which the row-wise elements of the matrix represent the performance or rating set of an alternative, and the column-wise elements express the performance of considered alternatives under the specific criterion (Hwang and Yoon 1981).

Following the framework of multi-attribute utility theory (MAUT), which is a model-based approach, multiple criteria formulation considers the aggregation of multiple criteria. It is used to represent the preference of a DM over bundles of goods either under conditions of certainty about the results of any potential choice or under conditions of uncertainty. With the assumptions of preference and utility independences, Keeney and Raiffa (1993) organized the formula into multiplicative and multi-linear forms. Attaching the assumption of additive independence, Von Winterfeldt and Edwards (1986) presented an additive form. MAUT tries to aggregate single value functions into a compact form that belongs to the compensatory model, under the assumption of criteria independence. The compensatory model allows trade-offs between criteria, meaning that an unfavorable value in one criterion can be offset by an advantage or favorable value in some other criteria (Hwang and Yoon 1981). Readers can reference the definitions of compensation and non-compensation in Bouyssou (1986).

Along the stream of multiple criteria formulation, there are a number of methods of multi-criteria decision making (MCDM) assuming that the values of the criteria are known with certainty (Korhonen et al. 1992). Hwang and Yoon (1981) classified them by three categories: (i) No preference information given as in Dominance, Maximin, and Maximax; (ii) Information on an attribute given as Conjunctive, Lexicographic, Simple Additive Weighting (SAW) or weighted sum methods, Analytic Hierarchy Process (AHP), Technique for Order of Preference by Similarity to Ideal Solution (TOPSIS), and ELimination Et Choice Translating REality (ELECTRE); and (iii) Information on actions or alternatives given as LINnear programming technique for Multidimensional Analysis of Preference (LINMAP). Guitouni and Martel (1998) sorted these models with four categories: (i) Elementary methods as Maximin and SAW; (ii) Single synthesizing criterion as AHP and TOPSIS; (iii) Outranking methods as ELECTRE and PROMETHEE (Preference Ranking Organization METHod for Enrichment of Evaluations) families; and (iv) Mixed methods as QUALIFLEX. Among them, the additive form with a linear function is the most commonly used MCDM, such as TOPSIS and AHP.

It is worth noting that elementary methods such as SAW are so simple that DMs might be reluctant to accept their results (Hwang and Yoon 1981). Thus we do not address this in detail. Any of the above methods could be applied to solve the problem. The MCDM process in practice should be: (i) identification of a problem or issue, (ii) problem structuring, (iii) model building, (iv) using the model to inform and for challenge thinking, and (v) developing an action plan (Belton and Stewart 2002). Compared to the problem-solving process in the previous section, it is obvious that the focus of this recommended process concentrates on managing problems.

The above formulation descriptions are within the scope of MAUT on gains or losses. Traditional utility is unable to fit gains and losses well in the real world, no matter how good MAUT is as a standard for multiple criteria formulation. Kahneman and Tversky (1979) observed people making judgments based on the potential gains or losses relative to their specific situation, the reference point, rather than in absolute terms; this is referred to as reference dependence. Their prospect theory provides deep insights into behavioral decision making under risk and offers a prosperous direction for multiple criteria formulation. Various types of uncertainties also arise when we intend to formulate the problems. As an illustration, vague data or uncertainty on estimating parameters could be formulated by fuzzy sets (Zadeh 1965). From a managerial aspect, we need to capture the critical uncertainties in the formulation and find a way to deal with the specific uncertainties (Olson and Wu 2010).

1.4 Alternatives

Alternatives are more available possibilities for choice. In making decisions, there are a couple of synonyms for actions, candidates, objects, options, projects, and variants, instead of alternatives. Other possible analogues could be key indices, successful factors, or forecasting trends in a specific problem. Alternatives can be investigated by multiple criteria models for ranking, selection, or sorting purposes, and they could be finite or infinite. Selecting a mayor from three candidates is an example of finite alternatives. In deriving an economical-order quantity, it is necessary to find a point with a minimum cost corresponding to the order quantity, which is an example of infinite alternatives. No matter if there are finite or infinite alternatives, alternatives can be classified into two sets: stable or evolutive and globalized or fragmented (Vincke 1992). Stable alternatives will not change or be modified during the decision-making process; otherwise, they are evolutive. Globalized alternatives are mutually exclusive, resulting in distinct alternatives for selection; however, fragmented alternatives involve combinations of several alternatives. For the final implementation, we might choose one from a set of mutually exclusive alternatives. In a complex situation, alternatives could be dependent, and they should be considered simultaneously to form a big set for the selection; e.g., research and development projects. Another situation that frequently happens in the capital budgeting process is one in which the selection has limited resources (Hartman 2007). In this case, the selection should be made under the condition of available resources. If an alternative costs more than the budget, then it will be screened out at the beginning of the process. If the budget is allowed, then a couple of alternatives could be included.

From a practical standpoint, dependent alternatives are unable to be evaluated directly. For instance, Fox et al. (1984) developed a more broad framework assessing expected present value, a benefit aspect, interactions among projects, and their impact on the formulation and solution of economic research and development project selection models. Gomes (1990) proposed a framework to model interdependencies among transportation system alternatives under multiple criteria, in which count interdependence factors are associated with alternatives through comparisons of their utilities with probabilities. This could be an interesting topic for future study.

1.5 Solutions

When we investigate the solution of the model, phase three of an operations research OR) study generally includes a couple of steps to obtain the solution. Solutions are the choices from DMs' preferences on alternatives. A few categories of solutions exist: feasible, infeasible, and optimal (Taha 2017). Most OR problems seek an optimal solution in a decision (variable) space whose variables are continuous or from infinite alternatives. When extending the problem to multi-objective decision making (MODM), there are multiple objectives to be satisfied, which usually conflict.

Hence, non-dominated solutions are the primary focus since optimal solutions are unobtainable in most situations unless these objectives have the same direction. Non-dominated solutions are defined if none of the objective functions or criterion vector can be improved in value without degrading some other objective values (Steuer 1986). The term efficiency is used when referring to points in the decision space. A point is efficient if its criterion vector is not dominated by the criterion vector of some other points in a feasible region. Pareto optimality, admissibility, non-dominated, and non-inferiority are all synonyms of efficiency. In the area of MCDM, all listed alternatives are feasible after a pre-defined screening process, and we aim to find a preferred or best one among alternatives.

To find a solution in any computational algorithm, the solving steps are to start the procedure, how to improve the existing solution, and how to find the final (optimal or near-optimal) solution. For example, as a simplex algorithm in linear programming, the procedure is defined as: (i) initiation; (ii) simplex direction; (iii) optimality; (iv) step size; and (v) new point and basis (Radin 2017). Steps (i) and (iii) check the solution for optimality and determine if a further step is needed if the optimal solution has not been reached. Step (iv) utilizes the minimum ratio test to determine a feasible solution or an unbounded case. However, there could be difficulty in finding a solution, and computational complexity is an issue. In MCDM, the solution procedure is much simpler due to finding a solution in a discrete space or a finite set of alternatives, commonly using arithmetical operations. Even though thousands of alternatives are evaluated, no computational complexity appears for ranking and selecting the best solution. With regard to sorting problems by MCDM, some works have discussed the contents of complexity analysis (Lourenzutti et al. 2017; Yu and Yang 2007), due to the fact that these operations are more complex than arithmetic operations. Moreover, the solution of the model is its output that reaches a recommended decision in MCDM. The model is realized by the procedure of solving problems, and there is no optimal solution involved in this regard (other than optimality of preference).

1.6 Validation and Verification

In multiple criteria modeling a solution is obtained after the formulation and solving procedure. It is obvious that a question will be raised on the reliability and robustness of the solution. How do we know that the solution indeed contributes toward solving our existing problem? This involves validation and verification (Daellenbach 1994). Originally, verification and validation are independent procedures that are used together for inspecting that a product, service, or system meets requirements and specifications, and that it fulfills its intended designed purpose (Wikipedia). Following the stream of operations research, in a MCDM procedure, validation concerns whether the model is a sufficiently valid representation of reality. With a simple sentence, are we building the model "right"? Verification focuses on if the model is mathematically correct and logically consistent, including its variables and parameters. In a similar fashion, we ask, are we building the "right model"? Both play an important role in confirming the formulation and the solution for which a decision will be made for further implementation or not (Daellenbach 1994). We observe that validation and verification sometimes are named validation for short (Hillier and Lieberman 2014), or external validation refers to validation, and internal validation represents verification (Daellenbach 1994).

Validation or model validation will determine whether or not the solution to the model makes sense in the context of the real problem. If this is not the case, then the model is not a faithful representation of reality and should be changed (Eiselt and Sandblom 2012). Since it is difficult to prove a model to be valid, an approximation may be acceptable. In other words, it should exhibit that the model is not wrong. In addition, validation reveals the creditability of the model, in which the essence properties can be demonstrated (Daellenbach 1994). If the validation fails, then the discrepancies allow the model builder to identify the aspects of the model, in which erroneous assumptions may have been made.

Gass (1991) classified validity into three parts: theoretical validity, data validity, and operational validity, which all belong to the evaluation process. The first one concerns whether the theory fits the problem environment, the second one focuses on whether the data are accurate, complete, impartial, and appropriate, and the last one pivots if the model diverges from users' perception of the real world. Whenever a simulation model is performed, the validation of the results is mandatory. This often means checking the computer code and performing statistical tests (Hillier and Lieberman 2014). In system engineering, validation concerns if the designed system could meet initially specified consumer requirements, and this cannot be fulfilled until the system is produced and functioning in an operational environment (Blanchard and Blyler 2016). Validation has been employed in many areas and is an essential step related to the model's usability.

Verification is the act or process to establish the truth, accuracy, or reality of something. A better way for verifying the model as being mathematically correct and logically consistent is by numerically checking the results by hand or spreadsheet for a sufficiently wide range of inputs and inspecting that each expression is dimensionally consistent as well (Daellenbach 1994). The correctness of all numerical constants should be verified. Gass (1991) also noted that verification attempts to confirm the computer-based model is running as intended. In software development, validation ensures that the software meets system requirements, that compatibility exists between software and hardware, and that appropriate quality specifications have been incorporated (Blanchard and Blyler 2016). During the validation phase of model development, animations can be used to point out anomalies in the movements of the transaction, but this advantage could be less restricted by human patience (Taha 2017).

Verification and validation are difficult tasks in multiple criteria formulation and solutions obtained, but not many published works address them. Wu and Tiao (2018) presented a comprehensive validation scheme for comparing six frequently-used MCDM methods. The scheme examined and compared the consistency of the alternative chosen by MCDM methods with DM's intuitive ideal alternative. The settings of control variables contained the number of alternatives, number of criteria, dataset distributions, and non-dominated dataset options; DM's preferences included various weight distributions and utility functions. Although results could not be compared with real outcomes, the analysis undeniably provides a guideline to check the consistency of a specific method. The study also dealt with sensitivity analysis. Sensitivity analysis is a tool to identify the sensitive parameters whose values cannot be changed without changing the optimal solution (Hillier and Lieberman 2014). In the real world, the parameter values used in the model normally are just estimates based on a prediction of future conditions. Thus, the parametric data may represent deliberate overestimates or underestimates to protect the interests of the estimators or to provide an imprecise estimation. The tool is popularly related to the solutions, and we could think of it as one part of verification. Aside from the line chart being used, spiderplots and tornado diagrams are common for illustrating the results of sensitivity analysis (Eschenbach 1992).

Another common tool is scenario planning, which makes assumptions about what the future will be and how one's business environment will change based on that future. The concept of scenario planning is being widely adopted as an important tool in strategic planning and analysis under conditions of uncertainty and imprecision (Comes et al. 2013; Stewart 1997). Islei et al. (1999) presented how judgmental models have been used to gain insights into a problem, in which a large group of experienced managers took part in a meeting to develop and discuss scenarios in pharmaceutical process engineering. Chermack (2004) stated that scenario planning is a tool for preventing the impact of four types of decision failure. Here, the tool of scenario planning could be considered as one part of the validation.

1.7 Summary and Discussions

In this chapter we have discussed some elements in the multiple criteria modeling process that are essential in making a good decision. However, it is hard to explain the real-world problem by its abstraction and conduct validation and verification to ensure the solution targets the problem. The modeling process is some kind of art, because the content depends on the problem itself and the stakeholders involved (Franco and Montibeller 2010). Nonetheless, we need to do our best to guarantee our modeling is in a right direction.

The selection of a proper MCDM method is another issue in multiple criteria formulation. Saaty and Ergu (2015) carried out method comparison and proposed 16 criteria that may be used to judge and evaluate these methods. The sixteen criteria include:

(i) simplicity of execution,
(ii) comprehensive structure of breadth and depth,
(iii) comprehensive structure consisting of merit substructures,
(iv) a logical, mathematical procedure,
(v) justification of the approach-justifiable axioms,
(vi) scales of measurement,
(vii) synthesis of judgments with merging functions,
(viii) ranking of tangibles,
(ix) generalization to the ranking of intangibles,
(x) rank preservation and reversal,
(xi) sensitivity analysis,
(xii) validation of decision problems,
(xiii) prediction of the outcome of decisions with intangibles (the capability to predict preferences),
(xiv) generalizability to dependence and feedback,
(xv) applicability to conflict resolution, and
(xvi) trustworthiness and validity of the approach, which provide a guideline for the selection.

Asgharizadeh et al. (2019) surveyed 17 MCDM methods and classified them by their output, containing simplicity, speed, memory, inputs, logic, quality, and rate of growth, through the fuzzy c-means technique. Two classes were identified according to the largest degree of membership. The first class includes SAW, TOPSIS, ORESTE, MAUT, VIKOR, COPRAS, SMART, MOORE, and ARAS, while the second covers ELECTRE I, PROMETHEE I, EVAMIX, REGIME, MAPPAC, TACTIC, ARGUS, and PACMAN. The results provide different aspects of these methods. Wątróbski et al. (2019) conducted a comprehensive survey on 64 MCDM methods, including some combinations of single methods, and presented a methodological and practical framework for selecting suitable MCDA methods for a particular decision situation. These contents are helpful for choosing appropriate methods from various features as required. Finally, we present a table for a comparison on the basic characteristics of the four prevalent methods for readers' reference (Table 1.1).

Table 1.1 Comparison of the characteristics of four common MCDM methods

No	Characteristic	AHP	EVAMIX	PROMETHEE	TOPSIS
1	Category	Information on the criterion with cardinal and ordinal information	Information on the criterion with cardinal and ordinal information	Information on the criterion with cardinal and ordinal information	Information on the criterion with cardinal information
2	Core process	(1) Make a pair-wise comparison to form the positive reciprocal matrix, using a 9-point scale for measurement, in a hierarchy (2) Calculate the eigenvalue of the matrices to obtain the priorities (3) Conduct matrix operations from the bottom up for the priorities of the hierarchy	(1) Use the distance function and pair-wise comparisons of the difference between the performance measures of alternatives to obtain the dominance scores (2) Apply the signum function to pre-process ordinal data on the qualitative criteria and merge them with the previous step by weights (3) Accumulate the normalized dominance scores for the ranking	(1) Make a pair-wise comparison on the differences of performance ratings of alternatives on the criteria, based on six types of preference functions with thresholds (2) Accumulate these preference indices to represent the dominance of alternatives by their inflows and outflows (3) Compare the relations of inflows and outflows of the alternatives for the ranking	(1) Use a distance function to measure the distances of alternatives from their PIS and NIS (2) Combine the two distances for the ranking
3	Criteria	Given	Given	Given	Given
4	Weight elicitation	Pair-wise comparison	Given	Given	Given
5	Consistency check	Provided	None	None	None
6	No. of criteria accommodated	7 ± 2 or by hierarchical decomposition	Many more	Many more	Many more
7	No. of alternatives accommodated	7 ± 2	Many more	Many more	Many more

(continued)

Table 1.1 (continued)

No	Characteristic	AHP	EVAMIX	PROMETHEE	TOPSIS
8	Independence and feedback	Capable by its extension, analytic network process (ANP)	None	None	None
9	Others	Compensatory model	Compensatory model	Compensatory model with outranking relations	Compensatory model

Note Please refer to Brans et al. (1984), Hwang and Yoon (1981), Saaty (1980, 1996), Shih et al. (2007), Shih (2016), and Voogd (1983)

References

Altshuller, G., Shapiro, R.: About technical creativity. Quest. Psychol **6**, 37–49 (1956) (in Russian)
Asgharizadeh, E., Yazdi, M. T., Balani, A. M.: An output-oriented classification of multiple attribute decision-making techniques based on fuzzy c-means clustering method. Int. Trans. Oper. Res. **26**, 2476–2493 (2019). https://doi.org/10.1111/itor.12449
Belton, V., Stewart, T.J.: Multiple Criteria Decision Analysis: An Integrated Approach. Kluwer, New York (2002). ISBN: 9781461514954
Blanchard, B.S., Blyler, J.E.: System Engineering Management, 5th edn. Wiley, Hoboken, New Jersey (2016). ISBN: 9781119225324
Bouyssou, D.: Some remarks on the notion of compensation in MCDM. Eur. J. Oper. Res. **26**, 150–160 (1986). https://doi.org/10.1016/0377-2217(86)90167-0
Bouyssou, D., Marchant, T., Pirlot, M., Tsoukiàs, A., Vincke, P.: Evaluation and Decision Models with Multiple Criteria: Stepping Stones for the Analyst. Springer, New York (2006). ISBN: 9780387310985
Brans, J.P., Vincke, P., Mareschal, B.: PROMETHEE: A new family of outranking methods in multicriteria analysis. In: Brans, J.P. (ed.) Operations Research '84, pp. 477–491. North-Holland, New York (1984). ISBN: 9780444875617
Chermack, T.J.: Improving decision-making with scenario planning. Futures **36**, 295–309 (2004). https://doi.org/10.1016/S0016-3287(03)00156-3
Clemen, R.T.: Making Hard Decisions: An Introduction to Decision Analysis, 2nd edn. Brooks/Cole Publishing, Brooks/Cole Publishing, Pacific Grove, California (1996). ISBN: 9780534260347
Comes, T., Hiete, M., Schultmann, F.: An approach to multi-criteria decision problems under severe uncertainty. J. Multi-Criteria Decis. Anal. **20**, 29–48 (2013). https://doi.org/10.1002/mcda.1487
Daellenbach, H.: Systems and Decision Making: A Management Science Approach. Wiley, West Sussex, Chichester (1994). ISBN: 9780471950943
Eiselt, H.A., Sandblom, C.L.: Operations Research: A Model-Based Approach, 2nd edn. Springer, Heidelberg (2012). ISBN: 9783642310539
Eschenbach, T.G.: Spiderplots versus tornado diagrams for sensitivity analysis. Interfaces **22**, 40–46 (1992). https://doi.org/10.1287/inte.22.6.40
Fox, G.E., Baker, N.R., Bryant, J.L.: Economic models for R and D project selection in the presence of project interactions. Manage. Sci. **30**, 890–902 (1984). https://doi.org/10.1287/mnsc.30.7.890
Franco, L.A., Montibeller, G. : Problem structuring for multicriteria decision analysis interventions. In: Cochran, J.J., Cox, Jr. L.A., Keskinocak, P., Kharoufeh, J.P., Smith, J.C. (eds.) Wiley Encyclopedia of Operations Research and Management Science. Wiley, New York (2010). ISBN: 9780470400630
Gass, S.I.: The many faces of OR. J. Oper. Res. Soc. **42**, 3–15 (1991). https://doi.org/10.1057/JORS.1991.2
Gomes, L.F.A.M.: Modelling interdependencies among urban transportation system alternatives within a multicriteria ranking framework. J. Adv. Transp. **24**, 77–85 (1990). https://doi.org/10.1002/atr.5670240107
Google: The ZMOT Handbook: Ways to Win Shoppers at the Zero Moment of Truth. Google, Mountain View, California (2012)
Guitouni, A., Martel, J.M.: Tentative guidelines to help choosing an appropriate MCDA method. Eur. J. Oper. Res. **109**, 501–521 (1998). https://doi.org/10.1016/S0377-2217(98)00073-3
Hartman, J.C.: Engineering Economy and the Decision-Making Process. Pearson, Upper Saddle, New Jersey (2007). ISBN: 9780131424012
Hillier, F., Lieberman, G.: Introduction to Operations Research, 10th edn. McGraw-Hill, New York (2014). ISBN: 9781259162985
Hwang, C.L., Yoon, K.: Multiple Attribute Decision Making: A State-of-the-Art Survey. Springer, Berlin (1981). ISBN: 9783540105589
Hyken, S.: The new moment of truth in business. Forbes (2016). https://www.forbes.com/sites/shephyken/2016/04/09/new-moment-of-truth-in-business/#bdce9a638d94

Ishikawa, K.: Guide to Quality Control. JUSE, Tokyo (1968). ISBN: 9789283310365
Islei, G., Lockett, G., Naudé, P.: Judgemental modelling as an aid to scenario planning and analysis. Omega **27**, 61–73 (1999). https://doi.org/10.1016/S0305-0483(98)00026-7
Jajuga, K., Sokolowski, A., Bock, H.H. (eds.): Classification, Clustering, and Data Analysis: Recent Advances and Applications. Springer, Heidelberg (2002). ISBN: 9783540436911
Jolliffe, I.T.: Principal Component Analysis, 2nd edn. Springer, New York (2002). ISBN: 9780387954424
Kahneman, D., Tversky, A.: Prospect theory: an analysis of decision under risk. Econometrica **47**, 263–292 (1979). https://doi.org/10.2307/1914185
Keeney, R.L., Raiffa, H.: Decisions with Multiple Objectives—Preferences and Value Tradeoffs, 2nd edn. Cambridge University Press, Cambridge (1993). ISBN: 9781139174084
Korhonen, P., Moskowitz, H., Wallenius, J.: Multiple criteria decision support—a review. Eur. J. Oper. Res. **63**, 361–375 (1992). https://doi.org/10.1016/0377-2217(92)90155-3
Lourenzutti, R., Krohling, R.A., Reformat, M.Z.: Choquet based TOPSIS and TODIM for dynamic and heterogeneous decision making with criteria interaction. Inf. Sci. **408**, 41–69 (2017). https://doi.org/10.1016/j.ins.2017.04.037
MacCrimmon, K.R.: Decisionmaking Among Multiple-Attribute Alternatives: A Survey and Consolidated Approach. RAND Corporation, Santa Monica, California (1968)
Olson, D.L., Wu, D.: Enterprise Risk Management Models. Springer, Heidelberg (2010). ISBN: 9783642114731
Osborn, A.F.: Applied Imagination: Principles and Procedures of Creative Thinking. Charles Scribner's Sons, New York (1953). ISBN: 9781447417101
Rardin, R.L.: Optimization in Operations Research, 2nd edn. Pearson, Hoboken, New Jersey (2017). ISBN: 9780134384559
Robertson, D.W., Jr.: A note on the classical origin of "circumstances" in the medieval confessional. Stud. Philol. **43**, 6–14 (1946)
Rowe, A.J., Boulgarides, J.D.: Managerial Decision Making. Macmillan, New York (1992). ISBN: 9780024041111
Saaty, T.L.: The Analytic Hierarchy Process. McGraw-Hill, New York (1980). ISBN: 9780070543713
Saaty, T.L.: Decision Making with Dependence and Feedback: The Analytic Network Process. RWS Publications, Pittsburgh (1996). ISBN: 9780962031793
Saaty, T.L., Ergu, D.: When is a decision-making method trustworthy? Criteria for evaluating multi-criteria decision-making methods. Int. J. Inf. Technol. Decis. Mak. **14**, 1171–1187 (2015). https://doi.org/10.1142/S021962201550025X
Saaty, T.L., Shih, H.S.: Structures in decision making: on the subjective geometry of hierarchies and networks. Eur. J. Oper. Res. **199**, 867–872 (2009). https://doi.org/10.1016/j.ejor.2009.01.064
Shih, H.S.: A mixed-data evaluation in group TOPSIS with differentiated decision power. Group Decis. Negot. **25**, 537–565 (2016). https://doi.org/10.1007/s10726-015-9451-9
Shih, H.S., Shyur, H.J., Lee, E.S.: An extension of TOPSIS for group decision making. Math. Comput. Model. **45**, 801–813 (2007). https://doi.org/10.1016/j.mcm.2006.03.023
Simon, H.A.: A behavioral model of rational choice. Quart. J. Econ. **69**, 99–118 (1955). https://doi.org/10.2307/1884852
Simon, H.A.: The New Science of Management Decision, 3rd edn. Prentice Hall, Englewood Cliffs, New Jersey (1977). ISBN: 9780136161363
Steuer, R.E.: Multiple Criteria Optimization: Theory, Computation, and Application. Wiley, New York (1986). ISBN: 9780471888468
Stewart, T. J. Scenario analysis and multicriteria decision making. In J. Clímaco (ed.), Multicriteria Analysis, 519–528. Springer, Heidelberg (1997). ISBN: 9783642606670.
Taha, H.A.: Operations Research: An Introduction, 10th edn. Pearson, Harlow, Essex (2017). ISBN: 9781292165547
Turban, E., Aronson, J.E., Liang, T.P., Sharda, R. : Decision Support and Business Intelligence Systems, 8th edn. Prentice Hall, Upper Saddle River, New Jersey (2006). ISBN: 9780131986602

Vincke, P.: Multicriteria Decision-Aid. Wiley, New York (1992). ISBN: 9780471931843
von Winterfeldt, D., Edwards, W.: Decision Analysis and Behavioral Research. Cambridge University Press, Cambridge (1986). ISBN: 9780521273046
Voogd. H.: Multicriteria Evaluation for Urban and Regional Planning. Pion, London (1983). ISBN: 9780850861068
Wang, J.J., Jing, Y.Y., Zhang, C.F., Zhao, J.H.: Review on multi-criteria decision analysis aid in sustainable energy decision-making. Renew. Sustain. Energy Rev. **13**, 2263–2278 (2009). https://doi.org/10.1016/j.rser.2009.06.021
Wątróbski, J., Jankowski, J., Ziemba, P., Karczmarczyk, A., Zioło, M.: Generalised framework for multi-criteria method selection. Omega **86**, 107–124 (2019). https://doi.org/10.1016/j.omega.2018.07.004
Wu, J.Z., Tiao, P.J.: A validation scheme for intelligent and effective multiple criteria decision-making. Appl. Soft Comput. **68**, 866–872 (2018). https://doi.org/10.1016/j.asoc.2017.04.054
Yoon, K.P., Hwang, C.L.: Multiple Attribute Decision Making: An Introduction. Sage, Thousand Oaks, California (1995). ISBN: 9780803954861
Yu, J., Yang, M.S.: A generalized fuzzy clustering regularization model with optimality tests and model complexity analysis. IEEE Trans. Fuzzy Syst. **15**, 904–915 (2007). https://doi.org/10.1109/TFUZZ.2006.889957
Zadeh, L.A.: Fuzzy sets. Inf. Control **8**, 338–353 (1965). https://doi.org/10.1016/S0019-9958(65)90241-X

Chapter 2
TOPSIS Basics

TOPSIS (Technique for Order Preference by Similarity to Ideal Solution) is a prominent distance-based MCDM/MADM (multi-criteria or multi-attribute decision making) technique. This chapter will review its origin followed by developments in the algorithm, along with a demonstration example. The intent is to show the basics of the method.

2.1 Origin, Benefits, and Drawbacks

TOPSIS is a useful technique to handle real world MCDM problems (Hwang and Yoon 1981). It helps decision maker(s) (DMs) conduct analysis, comparisons, and rankings of available alternatives when multiple criteria are involved. The basic idea of TOPSIS is rather straightforward. It originates from the concept of a displaced ideal point from which the solution has the shortest distance is identified (Belenson and Kapur 1973; Zeleny 1974). Hwang and Yoon (1981) further proposed that the ranking of alternatives is based on the shortest distance from the ideal solution or positive ideal solution (PIS) and the farthest distance from the negative-ideal solution (NIS) or anti-ideal solution or nadir in an n-dimensional Euclidean space. TOPSIS simultaneously considers the distances to both PIS and NIS, and a preference order is ranked on account of their relative closeness, which is a combination of these two distance measures. Although there are various versions, the core thinking of the technique is to believe that the distance function represents the DMs' preference or utility. Thus, the ranking of alternatives or variants can be made based on the combinations of the distances.

Kim et al. (1997) addressed three TOPSIS benefits: (i) a sound logic that expresses the rationale of human choice; (ii) a scalar value that considers both the best and worst alternatives simultaneously; and (iii) a simple computation process that can be easily programmed into a spreadsheet. Parkan and Wu (1997) deemed that TOPSIS has the

H.-S. Shih and D. Olson, *TOPSIS and its Extensions: A Distance-Based MCDM Approach*, Studies in Systems, Decision and Control 447,
https://doi.org/10.1007/978-3-031-09577-1_2

ability to identify the best alternative quickly. Zanakis et al. (1998) conducted simulations, finding that TOPSIS has the fewest rank reversals among the eight methods in the category. Olson (2004) considered it simple to use, and the only input beyond the raw data is the set of criteria weights. Shih et al. (2007) extended Kim et al.'s work and added the concept of a polyhedron to visualize performance measures of all alternatives on attributes, at least for any two dimensions. Roszkowska (2011) argued that the advantages of TOPSIS include simplicity, rationality, comprehensibility, good computational efficiency, and ability to measure the relative performance for each alternative in a simple mathematical form.

Huang and Li (2012) listed the advantages of TOPSIS as:

(i) a compromise can be efficiently obtained;
(ii) it uses logical thinking that represents the rationale of human choice;
(iii) it has a scalar value that expresses both the best and worst alternatives simultaneously;
(iv) it uses a comprehensible computation process that can be easily programmed into a spreadsheet;
(v) the performance measures of all alternatives on attributes can be visualized on a polyhedron (at least for any two dimensions); and
(vi) the results are easily explained to and accepted by DMs.

Zavadskas et al. (2016) listed eight positive features of TOPSIS. These benefits make TOPSIS an important MADM technique when compared with other related techniques such as Analytical Hierarchical Process (AHP) (Saaty 1980), ELimination Et Choix Traduisant la REalité (ELECTRE) (Benayoun et al. 1966), and Preference Ranking Organization METHod for Enrichment of Evaluation (PROMETHEE) (Brans et al. 1984), and the technology is pervasive around the world, especially in the Asia–Pacific region (Shih et al. 2007).

There also have been criticisms. Huang and Li (2012) selected the following three as the most critical:

(i) it does not provide a scheme for weight elicitation;
(ii) there is no unified normalization process; and
(iii) the judgment rule for the decision on the components of relative closeness is still open to question.

Kaliszewski and Podkopaev (2016) pointed out two basic issues for employing TOPSIS. Any ranking process consists of two parts: the cognitive part for the elicitation of the preference and the technical part for the calculation of the ranking. TOPSIS only realizes the technical part and the cognitive part is absent. The second one involves the weighting procedure. TOPSIS weights do not weight criteria directly, whereby the original idea to work with data of the same magnitude is lost. Zavadskas et al. (2016) listed four drawbacks:

(i) the normalized scale usually derives a narrow gap;
(ii) it never considers risk assessment;
(iii) it presents the problem of rank reversal; and

(iv) a contradiction may occur when a non-optimal alterative is replaced by a worse alternative.

From the viewpoint of ranking reversal or irregularity, Opricovic and Tzeng (2004) first illustrated the ranking irregularity for TOPSIS and initiated a large body of research. We shall discuss this issue in a later chapter.

2.2 TOPSIS Algorithm

The common algorithm of TOPSIS for ranking and selection includes the following seven steps (Hwang and Yoon 1981):

Step 1: Create a decision or evaluation matrix ***D***.

The matrix consists of m alternatives (A_1,..., A_m) and n criteria (X_1,..., X_n), with its element x_{ij}, where $i = 1, \ldots, m$ and $j = 1, \ldots, n$.

$$\boldsymbol{D} = \begin{array}{c} \\ A_1 \\ A_2 \\ \vdots \\ A_i \\ \vdots \\ A_m \end{array} \begin{array}{c} \begin{array}{cccccc} X_1 & X_2 & \cdots & X_j & \cdots & X_n \end{array} \\ \left[\begin{array}{cccccc} x_{11} & x_{12} & \cdots & x_{1j} & \cdots & x_{1n} \\ x_{21} & x_{22} & \cdots & x_{2j} & \cdots & x_{2n} \\ \vdots & \vdots & \cdots & \vdots & \cdots & \vdots \\ x_{i1} & x_{i2} & \cdots & x_{ij} & \cdots & x_{in} \\ \vdots & \vdots & \cdots & \vdots & \cdots & \vdots \\ x_{m1} & x_{m2} & \cdots & x_{mj} & \cdots & x_{mn} \end{array} \right]_{m \times n} \end{array} \tag{2.1}$$

Step 2: Construct the normalized decision matrix ***R***.

Matrix ***D*** is normalized to matrix ***R*** with m alternatives and n criteria, and its element r_{ij}, $r_{ij} = \frac{x_{ij}}{\sqrt{\sum_{i=1}^{m} x_{ij}^2}}$, where $i = 1, \ldots, m$ and $j = 1, \ldots, n$.

$$\boldsymbol{R} = \begin{array}{c} \\ A_1 \\ A_2 \\ \vdots \\ A_i \\ \vdots \\ A_m \end{array} \begin{array}{c} \begin{array}{cccccc} X_1 & X_2 & \cdots & X_j & \cdots & X_n \end{array} \\ \left[\begin{array}{cccccc} r_{11} & r_{12} & \cdots & r_{1j} & \cdots & r_{1n} \\ r_{21} & r_{22} & \cdots & r_{2j} & \cdots & r_{2n} \\ \vdots & \vdots & \cdots & \vdots & \cdots & \vdots \\ r_{i1} & r_{i2} & \cdots & r_{ij} & \cdots & r_{in} \\ \vdots & \vdots & \cdots & \vdots & \cdots & \vdots \\ r_{m1} & r_{m2} & \cdots & r_{mj} & \cdots & r_{mn} \end{array} \right]_{m \times n} \end{array} \tag{2.2}$$

Step 3: Construct the weighted normalized decision matrix ***V***.

A set of weights $\boldsymbol{w} = (w_1,\ldots, w_n)$, and $\sum_{j=1}^{n} w_j = 1$, where $w_j > 0, j = 1,\ldots, n$ is given to the corresponding criterion X_j, where $j = 1,\ldots, n$. The matrix $\boldsymbol{V} = [w_j r_{ij}]$ is calculated by multiplying the elements at each column of the matrix $\boldsymbol{R}$ by their associated weights $w_j, j = 1,\ldots, n$.

$$\boldsymbol{V} = \begin{array}{c} \\ A_1 \\ A_2 \\ \vdots \\ A_i \\ \vdots \\ A_m \end{array} \begin{array}{c} \begin{array}{ccccccc} X_1 & X_2 & \cdots & X_j & \cdots & X_n \end{array} \\ \begin{bmatrix} v_{11} & v_{12} & \cdots & v_{1j} & \cdots & v_{1n} \\ v_{21} & v_{22} & \cdots & v_{2j} & \cdots & v_{2n} \\ \vdots & \vdots & \cdots & \vdots & \cdots & \vdots \\ v_{i1} & v_{i2} & \cdots & v_{ij} & \cdots & v_{in} \\ \vdots & \vdots & \cdots & \vdots & \cdots & \vdots \\ v_{m1} & v_{m2} & \cdots & v_{mj} & \cdots & v_{mn} \end{bmatrix}_{m \times n} \end{array} \tag{2.3}$$

Step 4: Determine the positive ideal and negative-ideal solutions $\boldsymbol{V}^+$ (PIS) and $\boldsymbol{V}^-$ (NIS), respectively.

PIS and NIS are defined as $\boldsymbol{V}^+ = \{v_1^+, \cdots, v_n^+\} = \{(\max_i v_{ij} | j \in J), (\min_i v_{ij} | j \in J')\}$, and $\boldsymbol{V}^- = \{v_1^-, \cdots, r_n^-\} = \{(\min_i v_{ij} | j \in J), (\max_i v_{ij} | j \in J')\}$, respectively, where J is associated with the benefit criteria and J' is associated with the cost criteria, on criterion X_j, where $j = 1,\ldots, n$ for all alternatives A_i, $i = 1,\ldots, m$.

Step 5: Calculate the separation measure between alternative A_i, S_i^+ and S_i^-, and the ideal and the negative-ideal solutions, respectively.

The separation measure or distance between alternative A_i and the PIS can be measured by the n-dimensional Euclidean distance as follows.

$$S_i^+ = \sqrt{\sum\nolimits_{j=1}^{n} (v_{ij} - v_j^+)^2} \text{ for alternative } A_i, i = 1, \ldots, m. \tag{2.4}$$

In the same fashion, the separation measure or distance between alternative i and NIS can be illustrated as follows.

$$S_i^- = \sqrt{\sum\nolimits_{j=1}^{n} (v_{ij} - v_j^-)^2} \text{ for alternative} A_i, i = 1, \ldots, m. \tag{2.5}$$

Step 6: Calculate the relative closeness C_i^* of alternative A_i, $i = 1,\ldots, m$.

The relative closeness or ranking index of alternative A_i, $i = 1,\ldots, m$, is defined as:

$$C_i^* = \frac{S_i^-}{S_i^+ + S_i^-} \tag{2.6}$$

where $0 \leq C_i^* \leq 1$. The larger the index value is, the better is the performance of the alternative. The relative closeness is the judgment rule of the decision in TOPSIS. We observe that $C_i^* = 0$ if the alternative is the ideal solution and $C_i^* = 1$ if the alternative is the anti-ideal solution. In the real world, both are fictitious alternatives.

Step 7: Rank the preference order of all alternatives.

A set of alternatives A_i, $i = 1,\ldots, m$, can now be preference ranked according to the descending order of the value of C_i^*. In general, the selection should be the alternative with the highest value of the relative closeness.

We can conceptually illustrate a TOPSIS evaluation with a simple example shown in Fig. 2.1. Given two criteria, X_1 and X_2, with increasing preference (benefit criteria), all alternatives are inside the border of an ellipse. The relative closeness of alternative A_m is $C_m^* = \frac{S_m^-}{S_m^+ + S_m^-}$, and we can see its separation measures are measured by two lines S_m^+ and S_m^-, connecting the corresponding points PIS and NIS.

Note that the point PIS A^+ with the coordinated value $V^+ = \{v_1^+, v_2^+\}$ and the point NIS A^- with the value $V^- = \{v_1^-, v_2^-\}$.

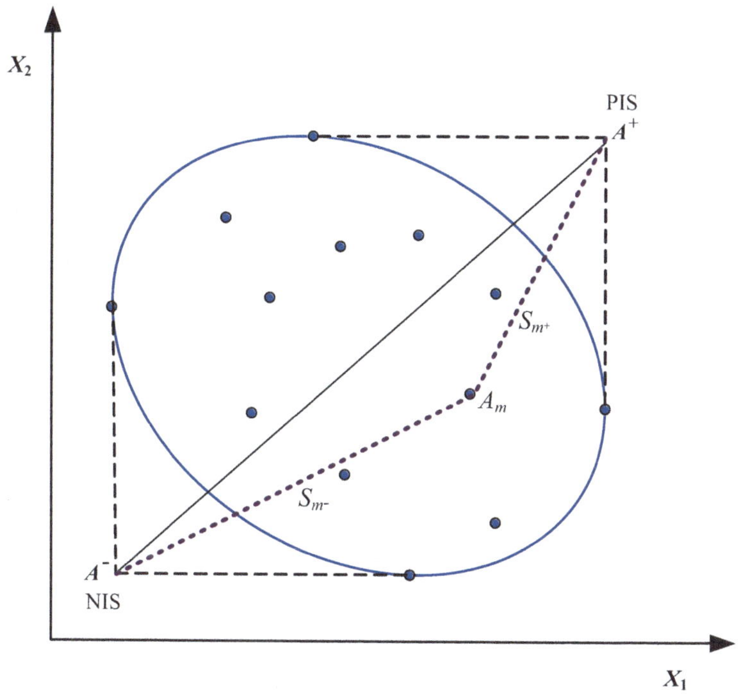

Fig. 2.1 Euclidean distances to PIS and NIS in a two-dimension space (Hwang and Yoon 1981)

2.3 TOPSIS Choice Behavior

TOPSIS ranks alternatives by their corresponding relative closeness, which is a combination of two separation measures. Hwang and Yoon (1981) investigated the choice behavior of TOPSIS and illustrated it by indifference curves with fixed values of their relative closeness. We reproduce some typical hyperbola curves in Fig. 2.2 under two benefit criteria X_1 and X_2 with equal preference or weight on the criteria in an Euclidean space. It shows that the curves with C_i^* greater than 0.5 are convex to the preference origin; i.e., PIS, which means the diminishing marginal rate of substitution on the two criteria; the curves with C_i^* less than 0.5 are concave to the preference origin. In the latter case, the behavior could be interpreted as a risk-prone behavior deriving from a pessimistic situation. In other words, when DMs realize their solution is closer to NIS than to PIS, they will be expected to take one criterion that has the best performance on one criterion and the other that holds the worst performance rather than to take an alternative that performs worse on two criteria. Yoon (1987) gave a supplemental example on a student selection. A committee might like to choose an applicant with a grade of A in one course and grade of F in the other course rather than one with grades of D in these two courses. The choice behavior will be in a reverse direction when C_i^* is greater than 0.5. For further discussion, this optimist and pessimistic behavior could be compared to the contents of the prospect theory (Kahneman and Tversky 1979).

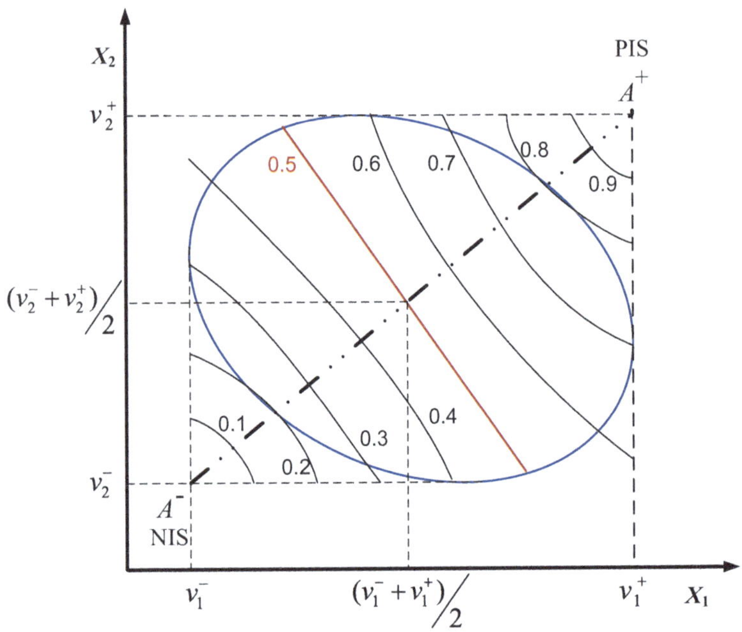

Fig. 2.2 Indifference curves in TOPSIS (Hwang and Yoon 1981)

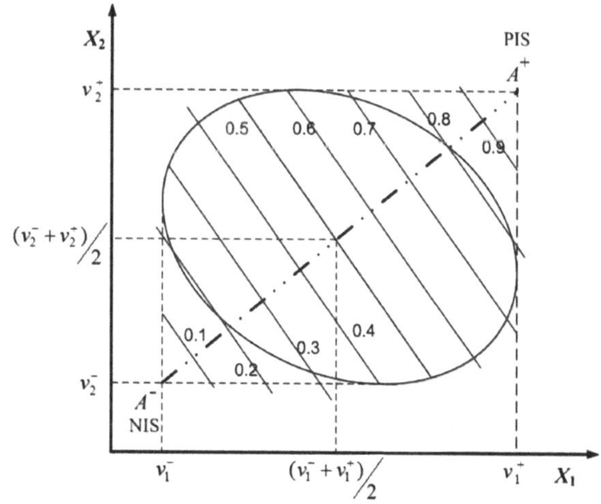

Fig. 2.3 Indifference curves in TOPSIS with Manhattan distance and equal weights (Lai et al. 1994)

Yoon (1987) also pointed out that indifference curves are a set of straight lines with slope of -1 under the Manhattan distance with equal weight on two criteria, and the curves under Tchebycheff or Chebyshev distance illustrated by connected segmented straight lines as well. Lai et al. (1994) further realized the contour of the Manhattan distance via parallel straight lines in a figure, and we modify it shown in Fig. 2.3 for a comparison purpose. Note that (Kaliszewski and Podkopaev 2016) also demonstrated similar contours with different weights given to two criteria. It is interesting to know the behavior when the value of C_i^* is equal to 0.5, which is at the center of the figure with red color and is perpendicular to the PIS-NIS line (a dot-dashed line). This is much like optimist and pessimistic free behavior, because it is a straight line without any tendency. Away from this line, the choice or judgment will have a distortion, no matter whether it is optimist or pessimistic. This distortion can be attributed to one major source of rank reversal in TOPSIS. From this viewpoint, the Manhattan distance with straight lines could supply a rational decision by its mathematical interpretation. Opricovic and Tzeng (2004) discussed an irregular situation on judging the alternatives' ranks and provided a figure to illustrate a case of the irregularity. Observe that (Hwang and Yoon 1981) also mentioned that TOPSIS is equivalent to the simple additive weighted (SAW) method if the process adopts the Manhattan distance, instead of the Euclidean distance, and in favor of PIS only.

2.4 TOPSIS Inputs

Unlike outranking methods, TOPSIS needs only two inputs for its evaluation: weights on criteria and the performance measures or ratings of all alternatives on the evaluated criteria. The weights on criteria are generally given by DMs or analysts in the beginning of the evaluation. Hwang and Yoon (1981) listed four methods of weight elicitation as a reference: Eigenvector method (Analytic Hierarchy Process), Weighted least square method, Entropy method, and LINMAP. As an example, Srdjevic et al. (2004) used entropy weights for TOPSIS in evaluating water management scenarios. We shall discuss the techniques of weight elicitation in great detail in the next chapter.

The second input infers that the performance measures or ratings of all alternatives on the evaluated criteria should be in quantitative form. This means that only quantitative data in the decision matrix, Eq. (2.1), can be calculated for further decision. It is common that the data could be qualitative, fuzzy, linguistic, interval, or other types of intensity of scale, which would require a data transformation. Hwang and Yoon (1981) proposed a 10-point scale for the transformation. For the benefit criteria, the values close to 10 are reserved for extremely favorable characteristics; i.e., very high, with values close to 5, and half, which is assigned for the average. On the low end of the scale, the values close to 1 are designated to be very low. The values of 7 and 3 stand for high and low, respectively. For the cost criteria, the values are arranged in a reverse direction. Figure 2.4 shows the scale transformation for qualitative data in Hwang and Yoon (1981).

Since the subjective assignment is rather straightforward, many other researchers follow in this fashion. Goh et al. (1996) utilized a 1- to 9-point scale for the level of achievement of the qualitative attribute for a robot selection problem. Parkan and Wu (1997) considered a 5-point Likert scale in a process selection problem. No matter which scale is adopted, even for a 0 to 1.0 normalized scale (Rao 2006), the designed scales or categories are often believed to illustrate an interval level of measurement. There should be a way to require a more objective evaluation to obtain the performance ratings, even with the qualitative criteria.

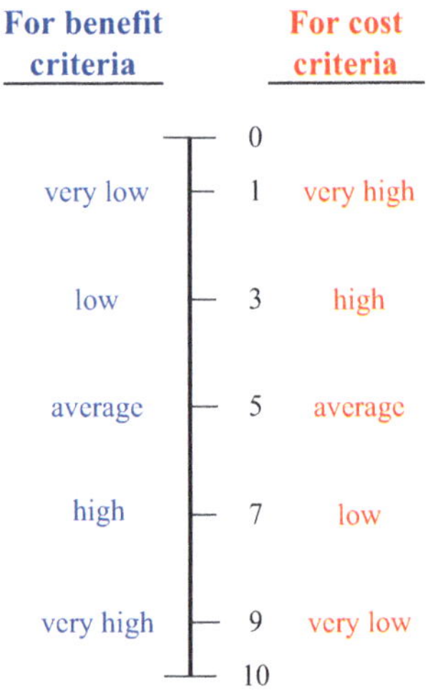

Fig. 2.4 Assignment of values for the qualitative scale (Hwang and Yoon 1981)

2.5 Current Developments

TOPSIS was designed for selection and ranking of a number of alternatives, and many applications have been published. Behzadian et al. (2012) collected 266 papers published in 103 scholarly journals since 2000. Nine major application areas are:

(i) supply chain management and logistics,
(ii) design, engineering, and manufacturing systems,
(iii) business and marketing management,
(iv) health, safety, and environment management,
(v) human resources management,
(vi) energy management,
(vii) chemical engineering,
(viii) water resources management, and
(ix) other topics.

Usually TOPSIS was combined with other techniques. The top ten techniques combined and compared with TOPSIS are the fuzzy set approach, group decision-making approach, AHP, entropy method, multi-objective optimization, genetic algorithms, ANP, Taguchi method, DEA, and simulation methods. Those papers were frequently published by the top ten journals in descending order: Expert Systems with Applications, International Journal of Advanced Manufacturing Technology, Mathematical and Computer Modelling, Applied Mathematical Modelling, Applied Soft Computing, Journal of Intelligent Manufacturing, International Journal of Production Research, Computers & Industrial Engineering, International Journal of Production Economics and European Journal of Operational Research. The top ten authors' nationalities are Taiwan, China, Iran, Turkey, India, U.S.A., Canada, Greece, UK, and Italy accordingly.

Zavadskas et al. (2016) examined 105 TOPSIS associated papers in the years 2000 to 2015 from Web of Science and Scopus databases, collected from 29 selected journals. They organized the works into two categories: based on developed and extended TOPSIS and presented and proposed TOPSIS extensions. They concluded that connections with group decision, fuzzy sets, and hybrid approaches are the recent developments. Zyoud and Fuchs-Hanusch (2017) conducted a bibliometric-based survey of AHP and TOPSIS methods. They collected 10,188 documents from Scopus for AHP and 2,412 TOPSIS documents and organized them by year, citation, geographic distribution, rank of top twenty most productive countries, rank of top twenty most productive journals, rank of top twenty prevalent areas of interests, rank of top twenty most cited articles, rank of top twenty most productive institutions, and rank of top twenty most prolific authors. The results showed the ten countries with the most TOPSIS articles are China, Iran, India, Taiwan, Turkey, United States, Lithuania, Malaysia, Canada, and United Kingdom, with Taiwanas the most influential country with 43.98 average citations per paper. The top ten most productive journals are listed as: Expert Systems with Applications, Applied Soft Computing Journal, International Journal of Advanced Manufacturing Technology, International

Journal of Production Research, Journal of Intelligent and Fuzzy Systems, Applied Mathematical Modelling, Technological and Economic Development of Economy, Mathematical Problems in Engineering, Computers and Industrial Engineering, Mathematical and Computer Modelling and Knowledge-Based Systems (tie). The top ten prevalent areas of interests are: engineering, computer science, mathematics, business (management and accounting), decision sciences, environmental science, energy, social sciences, earth and planetary sciences, and agricultural and biological sciences. In addition, they also provided two graphs on density visualization map and network visualization map in association with published research that utilized the TOPSIS method with 44 keywords, respectively. Papathanasiou and Ploskas (2018) described TOPSIS in the first chapter of their book and reviewed several traditional and fuzzy TOPSIS related papers. They concentrated on the implementation of Python.

In the fuzzy domain, Nădăban et al. (2016) provided a short survey on fuzzy TOPSIS. Palczewski and Sałabun (2019) also reviewed 25 papers of fuzzy TOPSIS applications in the last decade. Salih et al. (2019) surveyed fuzzy TOPSIS from 2007 to 2017, gathering 170 articles from the databases of IEEE Xplore, Web of Science, Elsevier ScienceDirect, and Springer and categorized them into four categories: type-1 fuzzy, type-2 fuzzy, others, and survey papers. Its applications comprised the ten major areas:

(i) business and marketing management,
(ii) supply chain management and logistics,
(iii) design, engineering, and manufacturing,
(iv) information technology and networking applications,
(v) human resources management,
(vi) transportation management,
(vii) health and safety management,
(viii) military applications,
(ix) energy management, and
(x) geographic information system applications.

They also reviewed challenges and recommendations for future research.

The above six survey papers and one chapter demonstrate different TOPSIS developments. Their classifications cannot be compared due to their non-identical coverage. Furthermore, social networks have been very influential on our society and business in recent years. TOPSIS research in this area has been reviewed by Zareie et al. (2018), Acuña-Soto et al. (2021).

2.6 Illustrative Example: A Recruitment and Selection Problem

A Taiwanese semiconductor corporation extended its business by selling its newly-developed light-emitting diode (LED) components to lighting and other manufacturers of LED products. The sales and marketing department of the corporation recruited sales engineers to promote the new business. After a screening process of applications, 15 candidates remained on the final list for an interview in the second stage (Shih 2016). The interview evaluation included four dimensions: knowledge (management and specialty), skills (communication, expressiveness, problem solving, and analytic), behavior (team-work and customer-oriented), and personal characteristics (vigor, confidence, compressive resistance, and integrity). Since these 12 items are difficult to evaluate by cardinal information, e.g., 0–100 scores, interval values were used with some limited levels for ease of evaluation. The human resources department (HR) introduced a new evaluating system to fit the company's needs. For ease of manipulation, HR ordinal data with five levels was used for interview evaluations, satisfying the firm's needs. Table 2.1 lists the evaluation sheet of the 15 candidates on the 12 items by the manager from the employing unit (EU).

Based on the scale transformation suggested by Hwang and Yoon (1981) shown in Fig. 2.2, the quantitative data of the evaluation are illustrated in Table 2.2.

The process first constructs the normalized decision matrix through using Eq. (2.2), vector normalization, and with equal weights, 1/12 for each criterion, and a weighted normalized decision matrix is obtainable. Based on the contents of weighted normalized decision matrix, we can determine the positive ideal and negative-ideal solutions. Applying Eqs. (2.4) and (2.5), the separation measures to PIS and NISfor alternative A_i, $i = 1,\ldots, m$, are then acquired, S_i^+ and S_i^-, respectively. Employing Eq. (2.6), the relative closeness C_i^* of alternative A_i, $i = 1,\ldots, m$, is derived and the rank order of all alternatives is obtained. The results of the separation measures, the relative closeness, and the rank of alternatives appear in Table 2.3. It is obvious that alternative A_{11} got first place and A_1 got last.

2.7 Summary and Discussions

This chapter delivers the basic algorithm of TOPSIS and supplies two inputs to the algorithm. Once the inputs are obtained from the DMs, the calculation process could be automatically implemented to arrive at the ranking results. An example illustrates the algorithm in detail. Although some criticisms of the algorithm mention that it does not directly consider DM's preferences, e.g., preference and indifference thresholds, it does provide a simple computing process and is useful for large datasets.

Table 2.1 Interview evaluation of 15 candidates of the case company

Candidate	Knowledge		Skill				Behavior		Personal characteristics			
	Management	Specialty	Communication	Expressiveness	Problem solving	analytic	Team-work	Customer-oriented	Vigor	Confidence	Compressive resistance	Integrity
	X_1	X_2	X_3	X_4	X_5	X_6	X_7	X_8	X_9	X_{10}	X_{11}	X_{12}
A_1	Poor	Average	Average	Average	Poor	Average	Average	Average	Good	Good	Average	Average
A_2	Good	Good	Average	Average	Good	Average	Average	Good	Average	Good	Average	Good
A_3	Poor	Average	Average	Good	Good	Good	Average	Average	Good	Average	Average	Average
A_4	Average	Good	Average	Average	Average	Average	Good	Average	Good	Good	Average	Good
A_5	Good	Good	Good	Good	Good	Average	Good	Good	Good	Good	Good	Good
A_6	Good	Good	Good	Average	Good	Good	Average	Good	Average	Good	Average	Average
A_7	Good	Good	Good	Average	Good	Good	Average	Good	Average	Good	Average	Average
A_8	Average	Average	Good	Good	Average	Average	Good	Good	Average	Good	Good	Average
A_9	Average	Good	Average	Average	Poor	Average	Good	Average	Average	Good	Average	Good
A_{10}	Average	Average	Good	Good	Good	Average	Average	Average	Good	Good	Average	Average
A_{11}	Good	Good	Good	Average	Good	Good	Good	Good	Good	Good	Good	Good
A_{12}	Average	Average	Average	Average	Average	Average	Average	Good	Average	Average	Average	Average
A_{13}	Very poor	Average	Good	Good	Average	Average	Average	Good	Good	Good	Good	Average
A_{14}	Average	Average	Average	Good	Good	Good	Average	Average	Good	Good	Average	Average
A_{15}	Average	Average	Average	Average	Good	Average	Average	Average	Average	Good	Average	Average

Table 2.2 The transformed evaluation of 15 candidates of the case company

Candidate	Criteria											
	X_1	X_2	X_3	X_4	X_5	X_6	X_7	X_8	X_9	X_{10}	X_{11}	X_{12}
A_1	3	5	5	5	3	5	5	5	7	7	5	5
A_2	7	7	5	5	7	5	5	7	5	7	5	7
A_3	3	5	5	7	7	7	5	5	7	5	5	5
A_4	5	7	5	5	5	5	7	5	7	7	5	7
A_5	7	7	7	7	7	5	7	7	7	7	7	7
A_6	7	7	7	5	7	7	5	7	5	7	5	5
A_7	7	7	7	5	7	7	5	7	5	7	5	5
A_8	5	5	7	7	5	5	7	7	5	7	7	5
A_9	5	7	5	5	3	5	7	5	5	7	5	7
A_{10}	5	5	7	7	7	5	5	5	7	7	5	5
A_{11}	7	7	7	5	7	7	7	7	7	7	7	7
A_{12}	5	5	5	5	5	5	5	7	5	5	5	5
A_{13}	1	5	7	7	5	5	5	7	7	7	7	5
A_{14}	5	5	5	7	7	7	5	5	7	7	5	5
A_{15}	5	5	5	5	7	5	5	5	5	7	5	5

Table 2.3 TOPSIS evaluation of 15 candidates of the case company

Candidate	S_i^+	S_i^-	C_i^*	Rank
A_1	0.0297	0.0124	0.2936	15
A_2	0.0180	0.0311	0.6328	5
A_3	0.0248	0.0205	0.4525	12
A_4	0.0195	0.0236	0.5468	9
A_5	0.0075	0.0351	0.8242	2
A_6	0.0165	0.0319	0.6583	3
A_7	0.0165	0.0319	0.6583	3
A_8	0.0180	0.0247	0.5783	6
A_9	0.0241	0.0214	0.4703	11
A_{10}	0.0198	0.0254	0.5625	8
A_{11}	0.0073	0.0352	0.8278	1
A_{12}	0.0242	0.0188	0.4366	13
A_{13}	0.0290	0.0188	0.3931	14
A_{14}	0.0196	0.0255	0.5650	7
A_{15}	0.0233	0.0222	0.4874	10

References

Acuña-Soto, C., Liern, V., Pérez-Gladish, B.: Normalization in TOPSIS-based approaches with data of different nature: application to the ranking of mathematical videos. Ann. Oper. Res. **296**, 541–569 (2021). https://doi.org/10.1007/s10479-018-2945-5

Behzadian, M., Otaghsara, S.K., Yazdani, M., Ignatius, J.: A state-of the-art survey of TOPSIS applications. Exp. Syst. Appl. **39**, 13051–13069 (2012)

Belenson, S.M., Kapur, K.C.: An algorithm for solving multicriterion linear programming problems with examples. Oper. Res. Q. **24**, 65–77 (1973). https://doi.org/10.1057/jors.1973.9. https://doi.org/10.1016/j.eswa.2012.05.056.

Benayoun, R., Roy, B., Sussman, B.: ELECTRE: Une méthode pour guider le choix en présence de points de vue multiples. Note de travail 49, SEMA-METRA International, Direction Scientifique (1966)

Brans, J.P., Vincke, P., Mareschal, B.: PROMETHEE: A new family of outranking methods in multicriteria analysis. In: Brans, J.P. (ed.) Operations Research '84, pp. 477–491. North-Holland, New York (1984). ISBN: 9780444875617

Goh, C.-H., Tung, Y.-C.A., Cheng, C.-H.: A revised weighted sum decision model for robot selection. Comput. Ind. Eng. **30**, 193–199 (1996). https://doi.org/10.1016/0360-8352(95)00167-0

Huang, Y.-S., Li, W.-H.: A study on aggregation of TOPSIS ideal solutions for group decision-making. Group Decis. Negot. **21**, 461–473 (2012). https://doi.org/10.1007/s10726-010-9218-2

Hwang, C.L., Yoon, K.: Multiple Attribute Decision Making: A State-of-the-Art Survey. Springer, Berlin (1981). ISBN: 9783540105589

Kahneman, D., Tversky, A.: Prospect theory: an analysis of decision under risk. Econometrica **47**, 263–292 (1979). https://doi.org/10.2307/1914185

Kaliszewski, I., Podkopaev, D.: Simple additive weighting—a metamodel for multiple criteria decision analysis methods. Exp. Syst. Appl. **54**, 155–161 (2016). https://doi.org/10.1016/j.eswa.2016.01.042

Kim, G., Park, C.S., Yoon, K.P.: Identifying investment opportunities for advanced manufacturing systems with comparative-integrated performance measurement. Int. J. Prod. Econ. **50**, 23–33 (1997). https://doi.org/10.1016/S0925-5273(97)00014-5

Lai, Y.J., Liu, T.Y., Hwang, C.L.: TOPSIS for MODM. Eur. J. Oper. Res. **76**, 486–500 (1994). https://doi.org/10.1016/0377-2217(94)90282-8

Nădăban, S., Dzitac, S., Dzitac, I.: Fuzzy TOPSIS: a general view. Procedia Comput. Sci. **91**, 823–831 (2016). https://doi.org/10.1016/j.procs.2016.07.088

Olson, D.L.: Comparison of weights in TOPSIS models. Math. Comput. Model. **40**, 721–727 (2004). https://doi.org/10.1016/j.mcm.2004.10.003

Opricovic, S., Tzeng, G.H.: Compromise solution by MCDM methods: a comparative analysis of VIKOR and TOPSIS. Eur. J. Oper. Res. **156**, 445–455 (2004). https://doi.org/10.1016/S0377-2217(03)00020-1

Palczewski, K., Sałabun, W.: The fuzzy TOPSIS applications in the last decade. Procedia Comput. Sci. **159**, 2294–2303 (2019). https://doi.org/10.1016/j.procs.2019.09.404

Papathanasiou, J., Ploskas, N.: Multiple Criteria Decision Aid: Methods, Examples and Python Implementations. Springer Nature, Cham (2018). ISBN: 9783319916460.

Parkan, C., Wu, M.L.: On the equivalence of operational performance measurement and multiple attribute decision making. Int. J. Prod. Res. **35**, 2963–2988 (1997). https://doi.org/10.1080/002075497194246

Rao, R.V.: Machinability evaluation of work materials using a combined multiple attribute decision-making method. Int. J. Adv. Manuf. Technol. **28**, 221–227 (2006). https://doi.org/10.1007/s00170-004-2348-4

Roszkowska, E.: Multi-criteria decision making models by applying the TOPSIS method to crisp and interval data. Multiple Criteria Decis Making **6**, 200–230 (2011)

Saaty, T.L.: The Analytic Hierarchy Process. McGraw-Hill, New York (1980). ISBN: 9780070543713

Salih, M.M., Zaidan, B.B., Zaidan, A.A., Ahmed, M.A.: Survey on fuzzy TOPSIS state-of-the-art between 2007 and 2017. Comput. Oper. Res. **104**, 207–227 (2019). https://doi.org/10.1016/j.cor.2018.12.019

Shih, H.S.: A mixed-data evaluation in group TOPSIS with differentiated decision power. Group Decis. Negot. **25**, 537–565 (2016). https://doi.org/10.1007/s10726-015-9451-9

Shih, H.S., Shyur, H.J., Lee, E.S.: An extension of TOPSIS for group decision making. Math. Comput. Model. **45**, 801–813 (2007). https://doi.org/10.1016/j.mcm.2006.03.023

Srdjevic, B., Medeiros, Y.D.P., Faria, A.S.: An objective multi-criteria evaluation of water management scenarios. Water Resour. Manage **18**, 35–54 (2004). https://doi.org/10.1023/B:WARM.0000015348.88832.52

Yoon, K.: A reconciliation among discrete compromise solutions. J. Oper. Res. Soc. **38**, 277–286 (1987). https://doi.org/10.1057/jors.1987.44

Zanakis, S.H., Solomon, A., Wishart, N., Dublish, S.: Multi-attribute decision making: a simulation comparison of selection methods. Eur. J. Oper. Res. **107**, 507–529 (1998). https://doi.org/10.1016/S0377-2217(97)00147-1

Zareie, A., Sheikhahmadi, A., Khamforoosh, K.: Influence maximization in social networks based on TOPSIS. Exp. Syst. Appl. **108**, 96–107 (2018). https://doi.org/10.1016/j.eswa.2018.05.001

Zavadskas, E.K., Mardani, A., Turskis, Z., Jusoh, A., Nor, K.M.: Development of TOPSIS method to solve complicated decision-making problems: an overview on developments from 2000 to 2015. Int. J. Inf. Technol. Decis. Mak. **15**, 645–682 (2016). https://doi.org/10.1142/S0219622016300019

Zeleny, M.: A concept of compromise solutions and the method of the displaced ideal. Comput. Oper. Res. **1**, 479–496 (1974). https://doi.org/10.1016/0305-0548(74)90064-1

Zyoud, S.H., Fuchs-Hanusch, D.: A bibliometric-based survey on AHP and TOPSIS techniques. Exp. Syst. Appl. **78**, 158–181 (2017). https://doi.org/10.1016/j.eswa.2017.02.016

Chapter 3
TOPSIS Variants

After four decades, numerous TOPSIS variants have been introduced. This chapter includes nine sections, one for each of the individual variants, as: normalization processes, weights on criteria, distance functions, selection of PIS/NIS, relative closeness formula, ordinal input, weights on separation measures, dependent criteria, and incremental analysis. All of them are reviewed with illustrative examples. These contents could provide deep insights into the TOPSIS algorithm.

3.1 Normalization Process

The normalization process of performance measures on each criterion are not needed in non-compensatory models or if no preference information is given (Hwang and Yoon 1981). The methods of maximin, maximax, and dominance are the examples. If preference is considered, then TOPSIS aggregates the performance measures of all criteria, commonly through weighting, and the normalization process is essential to eliminate warping from different scales of measures. When purchasing a car, the cost could be US$10,000–50,000, and comfort and reliability could be evaluated using 1–10 indices or 7-point Likert scales. Without normalization, the cost criterion would dominate comfort and reliability criteria if we keep the original scales. Yoon and Hwang (1995) stated that normalization aims at obtaining comparable scales for eliminating the scaling effect caused by different measurement units. To be able to compare different scales of criteria, we need to make these scales dimensionless, and this is done through the normalization process (Pavličić 2001). For ease of aggregation over multiple criteria, all scales are kept between 0 and 1.

From a mathematical viewpoint, an additive utility function for a finite set $\boldsymbol{x}$ with n criteria is defined as follows (Keeney and Raiffa 1976).

H.-S. Shih and D. Olson, *TOPSIS and its Extensions: A Distance-Based MCDM Approach*, Studies in Systems, Decision and Control 447,
https://doi.org/10.1007/978-3-031-09577-1_3

$$u(x) = \sum_{j=1}^{n} k_j u_j(x_j), \quad j = 1, \ldots, n \tag{3.1}$$

where k_j is a positive scaling constant and allows us to add the separate contributions to the total utility $\boldsymbol{u}(\boldsymbol{x})$, and $u_j(x_j)$ denotes the utilities on criteria or attributes j, $j = 1, \ldots, n$. Note that we only consider index j here, and there is no index for alternatives in a decision matrix due to its simplicity.

Naturally, $u_j(x_j)$ should be normalized within the interval [0, 1] so that the additive function could work in balance on all criteria. Hwang and Yoon (1981) defined the TOPSIS algorithm using vector normalization in Eq. (2.2). Their process simplifies the normalization calculation without differentiating the performance on benefit and cost criteria. The differentiation is made on the selection of PIS and NIS points, seeking the minimal values for PIS and maximal values for NIS on cost criteria, versus the opposite definition for benefit criteria. Yoon and Hwang (1995) further classified the criteria into three groups, benefit, cost, and non-monotonic, with different formulae. Pavličić (2001) considered three simple normalization forms—simple (one kind of linear form), linear, and vector—and concluded that the normalization processes have an impact on the inconsistent choice by ELECTRE, TOPSIS, and SAW methods. Shih et al. (2007) collected five types of normalized formulae; i.e., three linear, one vector, and one non-monotonic. Zavadskas and Turskis (2008) organized fives forms of normalization methods, including no normalization, and proposed a logarithmic normalization.

Jahan and Edwards (2015) reviewed the influence of normalization techniques on ranking by MCDM techniques. They considered 31 forms of normalization grouped into five categories: sum based (8 types), linear-ratio based (7 types), linear max-min (6 types), non-linear (4 types), and others (similar to monotonic functions with 6 types). Peldschus (2018) examined ten pairs of linear, concave, and convex transformation functions on a dimensionless interval; i.e., normalization within the interval [0, 1]. He also provided figures to illustrate their effects by examples. Nonetheless, non-linear functions such as square root and power functions over three are uncommon in real world applications. Acuña-Soto et al. (2021) utilized the concept of similarity of two fuzzy sets to normalize data based on the ideal solution of alternatives. They employed four forms for normalization: vector, linear (max-min), linear (max), and linear (sum).

One may observe that there are other related normalization procedures in the literature. For instance, Cables et al. (2016) considered a flexible reference point on an interval, instead of the maximal or minimal extreme values, and proposed the reference ideal method (RIM). They claimed that their proposed method has two benefits: being independent of the type of data and not risking rank reversal. In Table 3.1 we list eight typical forms of normalization processes with a reference utilizing that type.

The formulae in the second half of the table are less used in applications. To visualize the effect of normalization, we supply a simple example in Example 3.1 to show the effects of different normalization processes.

Table 3.1 Typical formulae in normalization processes

No.	Normalization type	Formulation	Note
1	Linear normalization I—max	$r_{ij} = \frac{x_{ij}}{x_j^*}, i = 1, ..., m; \ j = 1, ..., n$ for benefit criteria; $r_{ij} = \frac{x_j}{x_{ij}}$ or $r_{ij} = 1 - \frac{x_{ij}}{x_j^*}, i = 1, ..., m; \ j = 1, ..., n$ for cost criteria	Hwang and Yoon (1981)
2	Linear normalization II—max-min	$r_{ij} = \frac{x_{ij} - x_j}{x_j^* - x_j}, i = 1, ..., m; \ j = 1, ..., n$ for benefit criteria; $r_{ij} = \frac{x_j^* - x_{ij}}{x_j^* - x_j^{\sim}}, \ i = 1, \ldots, m; \ j = 1, \ldots, n$ for cost criteria	Hwang and Yoon (1981)
3	Linear normalization III—sum	$r_{ij} = \frac{x_{ij}}{\sum_{i=1}^{m} x_{ij}}, i = 1, ..., m; \ j = 1, ..., n$ for benefit criteria; $r_{ij} = \frac{1/x_{ij}}{\sum_{i=1}^{m} 1/x_{ij}}, i = 1, ..., m; \ j = 1, ..., n$ for cost criteria	Jee and Kang (2000) also called sum normalization
4	Vector normalization	$r_{ij} = \frac{x_{ij}}{\sqrt{\sum_{i=1}^{m} x_{ij}^2}}, i = 1, ..., m; \ j = 1, ..., n$ for benefit criteria; $r_{ij} = 1 - \frac{x_{ij}}{\sqrt{\sum_{i=1}^{m} x_{ij}^2}}$ or $r_{ij} = \frac{1/x_{ij}}{\sqrt{\sum_{i=1}^{m} (1/x_{ij})^2}}, i = 1, ..., m; \ j = 1, ..., n$ for cost criteria	Pavličić (2001)
5	Non-monotonic normalization	$r_{ij} = e^{-\frac{z^2}{2}}, z = \frac{(x_{ij} - x_j^0)}{\sigma_j}, i = 1, ..., m; j = 1, ..., n$ x_j^0 is the most favorable value, and σ_j is the standard deviation of performance rating on the *j*th criterion	Yoon and Hwang (1995)
6	Original distance to the normal reference range (ODNRR)	$r_{ij} = 1 - \frac{x_j^* - x_{ij}}{\sum_{i=1}^{m} x_j^* - x_{ij}}, i = 1, ..., m; \ j = 1, ..., n$ for benefit criteria; $r_{ij} = 1 - \frac{x_{ij} - x_j}{\sum_{i=1}^{m} x_{ij} - x_j}, i = 1, ..., m; j = 1, ..., n$ for cost criteria; $r_{ij} = \begin{Bmatrix} 1 - \frac{x_{ij} - (\mu + 1.96\sigma)}{\sum_{i=1}^{m} x_{ij} - (\mu + 1.96\sigma)}, & \text{if } x_{ij} > (\mu + 1.96\sigma) \\ 1 - \frac{(\mu + 1.96\sigma) - x_{ij}}{\sum_{i=1}^{m} (\mu + 1.96\sigma) - x_{ij}}, & \text{if } x_{ij} < (\mu + 1.96\sigma) \\ 1, \text{ if } x_{ij} \in [\mu \pm 1.96\sigma] & \end{Bmatrix}$ where $i = 1,\ldots,m; j = 1,\ldots,n$ for normal reference range (NRR) criteria	Zeng et al. (2013)
7	Non-linear normalization	$r_{ij} = \left(\frac{x_{ij}}{x_j^*}\right)^2, i = 1, ..., m; \ j = 1, ..., n$ for benefit criteria; $r_{ij} = \left(\frac{x_j}{x_{ij}}\right)^2, i = 1, ..., m; \ j = 1, ..., n$ for cost criteria	Zavadskas and Turskis (2008)

(continued)

Table 3.1 (continued)

No.	Normalization type	Formulation	Note
8	Logarithmic normalization	$r_{ij} = \frac{\ln(x_{ij})}{\ln(\prod_{i=1}^{m} x_{ij})}$, $i = 1, ..., m$; $j = 1, ..., n$ for benefit criteria; $r_{ij} = \frac{1 - \frac{\ln(x_{ij})}{\ln(\prod_{i=1}^{m} x_{ij})}}{m-1}$, $i = 1, ..., m$; $j = 1, ..., n$ for cost criteria	Zavadskas and Turskis (2008)

Notes 1. $x_j^* = \max_i\{x_{ij}\}$ and $x_j^- = \min_i\{x_{ij}\}$. 2. μ is the mean and σ is the standard deviation of the data, respectively

Example 3.1 A two-criteria problem.

There are ten alternatives, A_1 to A_{10}, with two benefit criteria, X_1 and X_2, to be normalized as in Table 3.2.

The normalized values of three linear, vector, non-linear, and logarithmic forms on criterion X_1 are listed in Table 3.3. The ratios on alternatives' performance under each form are also presented.

Note that the three linear and vector normalizations generate linear ratios with equally distributed performance; however, the non-linear and logarithmic normalization do not. The non-linear normalization produces greater ratios than other forms, spurring a recommendation to use them with caution. The normalized values of Alternative A_1 have zero value in linear normalization I and logarithmic normalization and thus marked "–" in their corresponding column cells due to a zero value on the first alternative. Their ratios are counted from the second alternative. For comparative purposes, we also show the normalized values of the above six forms in Fig. 3.1. Linear normalizations I and II and non-linear normalization provide the

Table 3.2 Performance of ten illustrative alternatives by two criteria

Alternative	Criteria	
	X_1	X_2
A_1	1	10
A_2	2	9
A_3	3	8
A_4	4	7
A_5	5	6
A_6	6	5
A_7	7	4
A_8	8	3
A_9	9	2
A_{10}	10	1

Table 3.3 Normalized values of the performance of ten alternatives on criterion X_1

	Linear I		Linear II		Linear III		Vector		Non-linear		Logarithmic	
	Value	Ratio	Value	Ratio	Value	Ratio	Value	Ratio	Value	Ratio	Value	Ratio
A_1	0.1	1	0.0000	–	0.0182	1	0.0510	1	0.01	1	0.0000	–
A_2	0.2	2	0.1111	1	0.0364	2	0.1019	2	0.04	4	0.0459	1.00
A_3	0.3	3	0.2222	2	0.0545	3	0.1529	3	0.09	9	0.0727	1.58
A_4	0.4	4	0.3333	3	0.0727	4	0.2039	4	0.16	16	0.0918	2.00
A_5	0.5	5	0.4444	4	0.0909	5	0.2548	5	0.25	25	0.1066	2.32
A_6	0.6	6	0.5556	5	0.1091	6	0.3058	6	0.36	36	0.1186	2.58
A_7	0.7	7	0.6667	6	0.1273	7	0.3568	7	0.49	49	0.1288	2.81
A_8	0.8	8	0.7778	7	0.1455	8	0.4077	8	0.64	64	0.1377	3.00
A_9	0.9	9	0.8889	8	0.1636	9	0.4587	9	0.81	81	0.1455	3.17
A_{10}	1.0	10	1.0000	9	0.1818	10	0.5096	10	1.00	100	0.1524	3.32

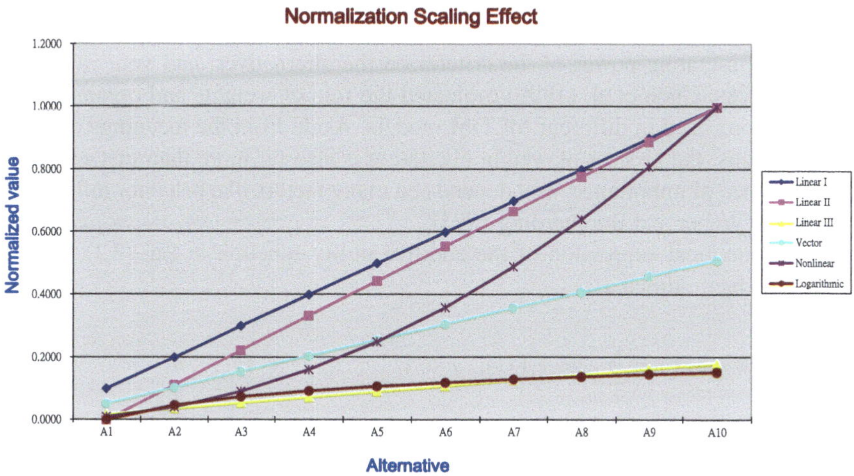

Fig. 3.1 Scaling effect of six normalization forms

maximal value 1 for the best performance alternative. Linear normalization III and vector and logarithmic normalizations generate few values for the best performance alternative, and their gaps seem to be condensed. In addition, the ratio by non-linear normalization is the highest. This means that its discriminating ability is higher than the others.

There are many discussions on the effect of the normalization process on ranking alternatives. Pavličić (2001) discussed the normalization effects on the results of ELECTRE, TOPSIS, and SAW methods, and three normalization forms were used for examining the rank reversal phenomenon of these three methods. Opricovic and Tzeng (2004)proposed VIKOR using the form of linear normalization II and claimed that VIKOR could avoid the weakness of TOPSIS. Kong (2011) also suggested that linear normalization II is better than vector normalization. Çelen (2014) considered four consistence conditions for the fuzzy TOPSIS normalization procedure and found that vector normalization generates the most consistent results. Chen (2019a) examined the combined effect of the normalization process with the use of entropy weights in TOPSIS. There are numerous works in dealing with the effects of the normalization process, and we will discuss them in later sections.

3.2 Weights for Aggregation

Many MCDM techniques, especially in compensatory models, require information about the relative importance of each criterion reflecting decision maker's relative preference (Yoon and Hwang 1995). A weight set is thus assigned to the relevant criterion. However, the meanings of weights involve more than just the interpretation of importance. For instance, the weight could reflect the rate of substitution among criteria, scaling factors used to convert measures into commensurate overall value, discriminating power of the criteria on the alternatives, and vote values in binary choices. Choo et al. (1999) evaluated the role of weights and organized 13 interpretations used in different MCDM models. Aside from the meanings of their interpretations, the process of weight elicitation is also far more than just catching the preference of importance, as it depends on many factors like behavior influences and others (Weber and Borcherding 1993).

A mathematical expression of the additive utility function in Eq. (3.1) with n criteria can be modified as:

$$\boldsymbol{u}(\boldsymbol{x}) = \sum_{j=1}^{n} w_j u_j(x_j), \quad j = 1, \ldots, n, j = 1, \ldots, n, \tag{3.2}$$

where w_j are the weights on criteria, and we keep $w_j \geq 0$ and $\sum_{j=1}^{n} w_j = 1, j = 1,\ldots,n$. The expression holds the property of convex combination, and thus every convex combination of elements of $\boldsymbol{x}$ also belongs to $\boldsymbol{x}$, or within the same decision space, for further decision. In addition, as described in Eq. (3.1), we only consider index j here for its simplicity.

In the early stages of multiple criteria evaluation, Eckenrode (1965) collected six methods for the judgments of experts concerning the relative value of sets of criteria based on psychological scaling methods and compared their reliability and time efficiency. The methods are ranking, rating, two types of partial paired comparisons, complete paired comparisons, and successive comparisons. The consistency among these methods was found to be very high. On the issue of time efficiency, the ranking method requires the least time to use. The other methods need more time with the method of successive comparison taking the longest. Cook and Stewart (1975) compared seven methods of subjective descriptions: dividing 100 points, rating on a 100-point scale, paired comparison, ratio, time influence, and two individual cue ratings. These methods perform reasonably well and yield more accurate predictions than equal or mean weights. Objective weights obtained by a multiple regression model has been discussed for predicting judgments; however, the goal of that approach is to accurately predict the outcomes for cues (Dawes and Corrigan 1974; Schmitts 1978).

The use of objective weights here is slightly different from the later classification of objective weighting techniques on criteria, with no information provided by the decision makers (DMs). Stillwell et al. (1981) compared six weight approximation techniques, including ratio, equal weights, rank sum, rank reciprocal, decision rule rank, and rank exponent, and concluded that rank weighting methods generally do better than equal weights. Schoemaker and Waid (1982) compared five approaches to include multiple linear and non-linear regression analyses and directly composed tradeoffs, eigenvalue, allocation of important points, and equal weight to determine weights in additive utility models for their prediction ability in college admission. They found that the performance of the equal weights is inferior to all other methods. Even so, Kirkwoodand Corner's experiments (1993) showed that rank-order information about weights is useful for small-size decision problems. Note that most of the above methods could be classified as subjective weighting techniques where criteria weights are derived from DMs' judgment.

Ordinal information is also valuable when making a decision if the cardinal information could not be provided or were uncertain. Barron and Barrett (1996) compared the decision quality of ranked attribute weights, including rank-sum, reciprocal of the ranks, rank exponents, and rank order centroid (ROC), and found ROC weights are more accurate than rank sum and rank reciprocal weights. The methods of weight by ranked order have the benefit of requiring only partial information and with much less input from DMs (Olson and Dorai 1992). They are also reliable as well (Larichev 1992). Many works assign cardinal weights generated from partial or order information (Alfares and Duffuaa 2009) and by ordered weighted averaging (OWA) operators (Yager 1988).

Objective weighting has been proposed for aggregating criteria values, because weights could be obtained from outcomes without asking DMs for their perceptions (Tzeng et al. 1998). In this approach, entropy has been widely used to investigate relationships between sets of data (Nijkamp 1977a). If the values for two alternatives on a criterion are the same, then the weight of the criterion will be zero and could thus be eliminated from the evaluation. This set of weighting methods includes criteria importance through intercriteria correlation (CRITIC) (Diakoulaki et al. 1995), constrained regression (García et al. 2010), and standard deviation (Jahan et al. 2012). In addition, the combinations of subjective and objective weighting techniques have been used. Ma et al. (1999) proposed an integrated approach to determine criterion weights through weighted least squares for subjective weights and objective weights measured by the distance to the ideal solution. In this approach a two-objective programming model is formulated to obtain the weight set. Jahan et al. (2012) heuristically proposed combinative weights, combining three types of weighting techniques: subjective, objective, and correlation by CRITIC. They then took the geometric mean of these three types in which the former two are any one of the class. Although many dozens methods have been developed, the choice of the weighting technique should depend on the essence of the problem.

Research has also examined weighting effects in TOPSIS. Olson (2004) compared weighting schemes, centroid, regression, and equal, with different distance metrics for TOPSIS in predicting winning baseball teams. He concluded that TOPSIS is not noticeably more accurate than direct weighting methods such as SMART, and all methods tested are quite close in accuracy except that equal weights are inferior. Kaliszewski and Podkopaev (2016) demonstrated that weights have no effect in the traditional TOPSIS process due to its inherent formulation. This observation leads us to re-examine the weighting procedure later.

Two articles reviewed details of weighting techniques. Riabacke et al. (2012) considered nine prominent weight elicitation methods—direct rating, point allocation, simple multi-attribute ranking technique (SMART), general swing weighting, trade-off methods, rank-order methods, eigenvalue weight (analytic hierarchy process, AHP), surrogate centroid criteria weight, and interval methods—and provided a summary of their assessment procedures. Zardari et al. (2015) extensively reviewed weighting methods and their effects on MCDM. They grouped these methods into three categories: subjective weighting, objective weighting and combination weighting; in which there are fifteen methods in the category of subjective weighting, seven methods in objective weighting, and five methods in combination weighting. These three classifications will be applied in our study. Some applications on weighting methods in combination with the MCDM techniques are also tabulated for reference. Furthermore, the choice of a weight set calls for attention. From the aspect of a rational decision, the choice should be non-dominated solutions or possess Pareto optimality. Kaliszewski (2018) analyzed the ranking of 50 Polish universities and found their best ranking would be one of the non-dominated solutions and the best one varied under different weight sets. This observation posts a critical question: Is any weight set appropriate for making a decision? There is no definite answer when we refer to the works of Choo et al. (1999) and Weber and Borcherding (1993).

Because of the large number of approaches of weight elicitation, only ten typical methods, under three types of weighting techniques, are selected in Table 3.4. We eliminate some subjective weighting methods, such as the ratio method, the point allocation method, Delphi method, etc., because they are subjective operations with less mathematical elements. In addition, the recently developed best-worst method (Rezaie 2015) can be categorized as a subjective weighting technique.

Example 3.2 Weights on criteria in the fighter selection problem (Hwang and Yoon 1981).

When a country decides to purchase a fleet of jet fighters from the United States, Pentagon officials provide four models with characteristics that should be considered on six criteria: maximum speed (X_1), ferry range (X_1), maximum payload (X_2),

Table 3.4 Typical methods for weight elicitation

No.	Classification	Elicitation technique	Description	Note
1	Subjective weighting	Equal weight	$w_j = \frac{1}{n}$, $j = 1, \ldots, n$ for all criteria	Dawes and Corrigan (1974)
2		Eigenvalue weight	$\boldsymbol{Aw} = \lambda \boldsymbol{w}$, where $\boldsymbol{A}$ is a pair-wise comparison $n \times n$ matrix whose elements are obtained from the comparison among n criteria, and $\boldsymbol{w}$ is the weight vector containing n corresponding weights. λ is the eigenvalue	(1) Saaty (1980) (2) Many other substitutive methods dealing with the pair-wise comparison matrix for weights; e.g., distance measure and least squares (Chu et al. 1979; Cook and Kress 1988)
3		Rank order centroid (ROC) weights	$w_j = \frac{1}{n} \sum_{k=i}^{n} \frac{1}{k}$, $j = 1, \ldots, n$ for all criteria under the jth order	(1) Barron and Barret (1996) (2) Only needs partial information from DMs
4		Ordered weighted average (OWA) weight	Quantifier weight for OWA operator $w_j = Q(\frac{j}{n}) - Q(\frac{j-1}{n})$, $j = 1, \ldots, n$ for all criteria under the jth order Regular increasing monotone (RIM) function $Q(r) = r^{\alpha}, \alpha \geq 0$	Yager (1996)
5	Objective weighting	Entropy weight I	(1) Normalize decision matrix $p_{ij} = \frac{x_{ij}}{\sum_{i=1}^{m} x_{ij}}$, $i = 1, \ldots, m$; $j = 1, \ldots, n$ (2) Calculate entropy $E_j = \left[\sum_{i=1}^{m} p_{ij} \ln(p_{ij})\right] / \ln(m)$, $j = 1, \ldots, n$ (3) Obtain entropy weight $w_j = \frac{1-E_j}{\sum_{j=1}^{n}(1-E_j)}$ for all criteria	Nijkamp (1977b)

(continued)

Table 3.4 (continued)

No.	Classification	Elicitation technique	Description	Note
6		Standard deviation weight	$w_j = \frac{\sigma_j}{\sum_{j=1}^{m} \sigma_j}$ for all criteria	Jahan et al. (2012)
7		Criteria importance through intercriteria correlation (CRITIC)	(1) Normalize decision matrix $\frac{x_{ij}-x_j}{x_j^*-x_j}, i = 1, ..., m;\ j = 1, ..., n$ (2) Establish a symmetric matrix with correction coefficient $r_{jl}, j, l = 1,...,n$ the performance among criteria, and calculate the measure of the conflict of criterion j with respect to the other criteria $m_j = \sum_{l=1}^{n} (1 - r_{jl})$ (3) Calculate the composite measure C_j by multiplying the measure of the conflict m_j by the standard deviation σ_j of criterion j $C_j = \sigma_j \sum_{l=1}^{n} (1 - r_{jl})$ (4) Obtain the weight by a normalization process $w_j = \frac{C_j}{\sum_{j=1}^{n} C_j}$ for all criteria	(1) Diakoulaki et al. (1995) (2) Consider the correction among the performance on criteria (3) Use linear normalization II—max–min for managing possible negative values

(continued)

Table 3.4 (continued)

No.	Classification	Elicitation technique	Description	Note
8		Constrained regression weight	Use extended goal programming (GP) model for estimating the weights of the performance measures and decision matrix, where weighted GP and Minmax GP are two extreme cases	(1) García et al. (2010) (2) Use linear normalization II—max–min for managing possible negative values
9	Combination weighting	Entropy weight II	Incorporation with given weight $\overline{w_j}$ from DMs $w_j = \frac{\overline{w_j}(1-E_j)}{\sum_{j=1}^{n} \overline{w_j}(1-E_j)}, j = 1,\ldots, n$ for all criteria	Hwang and Yoon (1981)
10		Combinative weight	Consider at least three types of the weights and make a geometric mean on them $w_j = \frac{\sqrt[3]{w_{j1} w_{j2} w_{j3}}}{\sum_{j=1}^{n} \sqrt[3]{w_{j1} w_{j2} w_{j3}}}, j = 1, \ldots, n$ for all criteria W_{j1}, W_{j2}, and W_{j3} represent subjective, objective, and correlation weights, respectively	(1) Jahan et al. (2012) (2) Subjective and objective weights could be chosen from one of the above two categories, and the correlation weight is CRITIC

Table 3.5 Decision table for the fighter selection problem with qualitative criteria

Alternative	Criteria					
	Maximum speed (Mach)	Ferry range (NM)	Maximum payload (pounds)	Purchasing cost ($ × 10^6)	Reliability (high-low)	Maneuverability (high-low)
	X_1	X_2	X_3	X_4	X_5	X_6
A_1	2.0	1,500	20,000	5.5	Average	Very high
A_2	2.5	2,700	18,000	6.5	Low	Average
A_3	1.8	2,000	21,000	4.5	High	High
A_4	2.2	1,800	20,000	5.0	Average	Average

Table 3.6 Decision table for the fighter selection problem with all quantitative criteria

Alternative	Criteria					
	Maximum speed (Mach)	Ferry range (NM)	Maximum payload (pounds)	Purchasing cost ($ × 10^6)	Reliability (high-low)	Maneuverability (high-low)
	X_1	X_2	X_3	X_4	X_5	X_6
A_1	2.0	1,500	20,000	5.5	5	9
A_2	2.5	2,700	18,000	6.5	3	5
A_3	1.8	2,000	21,000	4.5	7	7
A_4	2.2	1,800	20,000	5.0	5	5

acquisition cost (X_4), reliability (X_5), and maneuverability (X_6). The six measures for each alternative are shown in Table 3.5.

By using the assignment of values for the qualitative scale in Fig. 2.2 (Hwang and Yoon 1981), we obtain the decision table with all quantitative criteria for further processing in Table 3.6.

In the original example, the relative importance of the six criteria is given by the DMs as the weight set $\boldsymbol{w} = (w_1, w_2, w_3, w_4, w_5, w_6) = (0.2, 0.1, 0.1, 0.1, 0.2, 0.3)$. The assumption is based on the subjective weight, which represents the preference of the DMs. However, there are many techniques to elicit the criteria weights as illustrated in Table 3.4. We will calculate the corresponding weights that keep the same importance order as $w_6 \succ w_1 \approx w_5 \succ w_2 \approx w_3 \approx w_4$. In addition, some objective weights will obtain their weight from the performance data from Table 3.6. The resulting weights with some fixed parameters under ten methods are listed in Table 3.7. Note that the original given weight set could be obtained from any subjective weighting technique, such as ratio or point allocation methods. When comparing the weight sets from different methods in the example, we find that the eigenvalue and ordinal methods generally assign a large weight on the first important criterion. The constraint regression weight could assign a zero value to some unimportant criteria, which means we should take a very cautious step in employing the method.

Table 3.7 Weights of six criteria obtained from ten typical methods

No.	Classification	Elicitation technique	Weight set on criteria						Note
			Maximum speed	Ferry range	Maximum payload	Purchasing cost	Reliability	Maneuverability	
			w_1	w_2	w_3	w_4	w_5	w_6	
0	Subjective weighting	Original given weight	0.2	0.1	0.1	0.1	0.2	0.3	From Hwang and Yoon (1981) as a reference $\overline{w_j}$, where $j = 1,\ldots,6$
1	Subjective weighting	Equal weight	1/6	1/6	1/6	1/6	1/6	1/6	
2		Eigenvalue weight	0.2857	0.1429	0.1429	0.1429	0.2857	0.4286	Based on the ratio of original given weight
3		ROC weight	0.1944	0.0833	0.0833	0.0833	0.1944	0.3611	Based on the order of original given weight
4		OWA weight	0.1472	0.0975	0.0975	0.0975	0.1472	0.4131	(1) Based on the order of original given weight (2) $\alpha = 0.5$ in the RIM function as an optimistic attitude
5	Objective weighting	Entropy weight I	0.0642	0.2061	0.0134	0.0813	0.3587	0.2763	Based on the information from the decision matrix
6		Standard deviation weight	0.1650	0.1644	0.1623	0.1652	0.1579	0.1852	

(continued)

Table 3.7 (continued)

No.	Classification	Elicitation technique	Weight set on criteria						Note
			Maximum speed	Ferry range	Maximum payload	Purchasing cost	Reliability	Maneuverability	
			w_1	w_2	w_3	w_4	w_5	w_6	
7		CRITIC weight	0.2355	0.2101	0.1268	0.1304	0.1207	0.1765	
8		Constrained regression weight	0.2530	0.0911	0	0.4636	0	0.1923	(1) By the extended GP model where $\lambda = 0.5$ (2) The weight set by the weighted GP model ($\lambda = 0$) is [0.3309, 0.0731, 0.4707, 0, 0, 0.1252] and by the minmax GP model ($\lambda = 1$) is [0, 0, 0, 0.3333, 0.6667, 0]
9	Combination weighting	Entropy weight II	0.0650	0.1043	0.0068	0.0412	0.3631	0.4196	Incorporation with original given weight

(continued)

Table 3.7 (continued)

No.	Classification	Elicitation technique	Weight set on criteria						Note
			Maximum speed	Ferry range	Maximum payload	Purchasing cost	Reliability	Maneuverability	
			w_1	w_2	w_3	w_4	w_5	w_6	
10		Combinative weight	0.1580	0.1781	0.0606	0.1115	0.2245	0.2673	One combination of the original given weight, entropy weight, and CRITIC weight

3.3 Distance Functions

A distance function, whether distance or metric, is a function that gives a distance between each pair of point elements of a set. It is a numerical measurement of how far apart objects or points are. The term is commonly used in people's daily life, physics, mathematics, geographic survey, engineering, etc. From the mathematical aspect, it is a way of describing what it means for elements of some space to be close to or far away from each other. Though the distance could be a non-numerical measurement in psychology and social sciences, we like to present it by a value so that a comparison could be made in making a decision.

In mathematics, a distance function on a given set $(x, y, z) \in \boldsymbol{X}$ is a function d: $\boldsymbol{X} \times \boldsymbol{X} \rightarrow \boldsymbol{R}$, where $\boldsymbol{R}$ stands for the set of real numbers that satisfies the following conditions (Wikipedia):

(i) $d(x, y) \geq 0$, and $d(x, y) = 0$ if and only if $x = y$. (Distance is positive between two different points and is zero precisely from a point to itself.)
(ii) It is symmetric: $d(x, y) = d(y, x)$. (The distance between x and y is the same in either direction.)
(iii) It satisfies the triangle inequality: $d(x, z) \leq d(x, y) + d(y, z)$. (The distance between two points is the shortest distance along any path).

Such a distance function is known as a metric. Together with the set, it makes up a metric space.

The distance function has been widely applied to the area of MCDM for a long time. The concept of distance from goals or target values (i.e., deviations) has been used in goal programming, in which all deviations should be minimized (Charnes et al. 1955). Compromise programming identifies an ideal point as the DM's preferred point and calculates the solution on the minimized distance. The best-compromise solution is the nearest solution to the ideal point (Yu 1973; Zeleny 1974). In multi-objective decision making or multi-objective programming/optimization, the distance-function-based approach and achievement function approach take advantage of the same concept for the evaluation (Wiecek et al. 2016).

Distance measurement is a big issue in TOPSIS and has an impact on choice behavior as addressed in Sect. 2.3. The traditional TOPSIS solely considers the Euclidean distance for the measurement. Shih et al. (2007) exploited the contents of Steuer (1986) to deliver a general form, named as Minkowski L_p metrics and weighted L_p metrics. The Minkowski metric is a metric in a vector space that includes the Manhattan or city block distance $p = 1$, the Euclidean distance $p = 2$, and the Tchebycheff or Chebyshev distance $p = \infty$. These three are widely accepted for assessment and comparison; e.g., Jones and Mardle (2004), Lai et al. (1994), Shih (2016), and Yoon (1987). Please check Table 3.8 for a detailed explanation of the formulae. In addition, Chang et al. (2010) utilized $p = 1.5$ for mutual fund performance evaluation. Despite the fact that Steuer (1986) provided the contours or loci of Minkowski L_p metrics and weighted Minkowski L_p metrics, more complete contours could be found in Wikipedia under the term Minkowski distance.

Table 3.8 Commonly used distances in TOPSIS

No.	Distance measure	Note
1	Minkowski L_p metrics for $\boldsymbol{x}$ and $\boldsymbol{y} \in \boldsymbol{R}$ $L_p = \left[\sum_{j=1}^{n} \lvert x_j - y_j \rvert^p\right]^{1/p}$, where $p \geq 1$ on n dimensions (1) Manhattan distance $p = 1$, (2) Euclidean distance $p = 2$, (3) Tchebycheff distance $p = \infty$, Other positive values are feasible	(1) Steuer (1986) and Wikipedia (2) $L_{p=\infty} = \lim_{p\to\infty}\left(\sum_{j=1}^{n} \lvert x_j - y_j \rvert^p\right)^{1/p}$ $= \max_{j=1}^{n} \lvert x_j - y_j \rvert$, which emphasizes one of the largest deviations among n dimensions
2	Weighted Minkowski L_p metrics for $\boldsymbol{x}$ and $\boldsymbol{y}$ $\in \boldsymbol{R}$ $L_p = \left[\sum_{j=1}^{n} w_j \lvert x_j - y_j \rvert^p\right]^{1/p}$, where $p \geq$ 1 on n dimensions where w_j is the weight given to the jth dimension, $j = 1,\ldots, n$, and the p value is the defined to be the same as the above formula	Steuer (1986)

Chang et al. (2010) also provided distance curves between two points $(\boldsymbol{x}, \boldsymbol{y})$ under two dimensions, criteria $\boldsymbol{X}_1$ and $\boldsymbol{X}_2$ here. An expanded figure appears in Fig. 3.2. The physical property of the distance is notable in that its distance becomes less as the value of p increases; i.e., $L_{p=1} > L_{p=1.5} > L_{p=2} > \ldots > L_{p=\infty}$. In the figure, the Tchebycheff distance $L_{p=\infty} = \max\{\lvert x_1 - y_1\rvert, \lvert x_2 - y_2\rvert\} = \lvert x_1 - y_1\rvert$, marked by a yellow solid line for comparison. After the property is supplied, one might call into question about the impact of different values of p on a decision. Remark that Yoon (1987) concluded that the credibility of a distance function decreases when the value of p increases. This observation could be a guideline in choosing a proper p value if DMs have no other preference.

It is worth noting that the S-shape value function of the prospect theory (Kahneman and Tversky 1979) could be applied to the MCDM methods. Gomes and Lima(1991, 1992) first introduced the TODIM (a Portuguese acronym for Interactive and Multi-criteria Decision Making) method in which the value functions are pair-wise compared and referred to for gains and losses. The S-shape function could be viewed as a distance measure, as a non-linear form with two directions. For instance, Fan et al. (2013) first set aspiration levels as the reference points, calculated the gains and losses of alternatives' performances to form two decision matrices, and combined these two for the decision. Yin and Shyur (2018) directly utilized the S-shape function to gauge the distances from two references, extreme positive/negative feasible solutions, and then employed the relative closeness for the selection purpose. Through the experiment demonstrated the proposals to be effective, the election of the parameters and the reference points is still a challenge.

Some other distance functions have been applied to TOPSIS. The Mahalanobis distance is a measure of the distance between a point $\boldsymbol{P}$ and a distribution $\boldsymbol{D}$ (Mahalanobis 1936). It is considered to be a generalization of the Euclidean distance with a

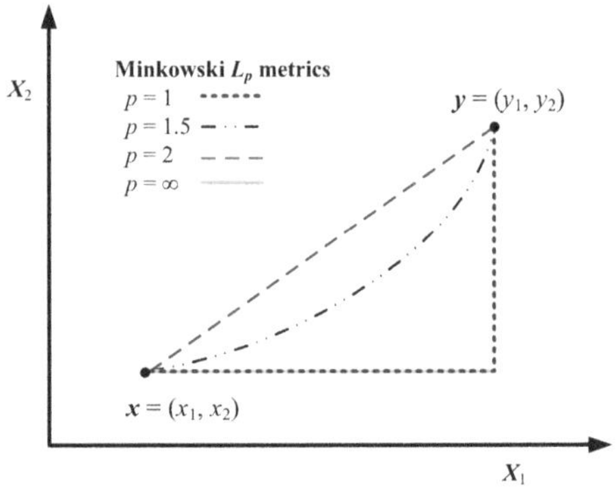

Fig. 3.2 Distance curves between two points (Chang et al. 2010)

correlation effect on unit length. Chang et al. (2010) utilized it for assessing mutual fund performance and found the result is rather consistent. The Hellinger distance is used to quantify the similarity between two probability distributions in the area of probability and statistics (Hellinger 1909). It is defined in terms of the Hellinger integral. Lourenzutti and Krohling (2014) applied the Hellinger distance to manage probability distribution information directly for TOPSIS and TODIM methods. They realized that the proposed approach is simple to handle such kind of information through experiments. In information theory, the Hamming distance measures the minimum number of substitutions required to change one string into the other or the minimum number of errors that could have transformed one string into the other (Hamming 1950). Izadikhah (2009) made use of the Hamming distance for calculating the distance between two triangular fuzzy numbers for fuzzy TOPSIS. He also devoted the procedure to a nurse recruitment application in a hospital. Distance measure is one core process in the TOPSIS algorithm, and we have listed some to the best of our knowledge from the literature. Most variants deal with the correlation among the performances on different criteria, and we will discuss them in Chap. 7.

We use a fighter selection model to demonstrate which different p values of Minkowski L_p metrics have impacts on the final ranks of alternatives.

Example 3.3 The fighter selection problem (Hwang and Yoon 1981).

Following the data given in Table 3.6, using the weight set w = (0.2, 0.1, 0.1, 0.1, 0.2, 0.3 (Hwang and Yoon 1981), we calculate separation measures and relative closeness under the values of metrics p = 1, 1.5, 2, and ∞, respectively. Their results are shown in Table 3.9.

Based on the values of their relative closeness, the alternatives could be ranked. For the metric $p = 1$, the ranking order is $A_3 \succ A_1 \succ A_4 \succ A_2$, and the order of the rest of the metrics is stable, which is $A_1 \succ A_3 \succ A_4 \succ A_2$. Thus, there are slightly different ranking orders. In later chapters, we will discuss the reasons for rank reversal. Note that the example verifies the phenomenon that the value of the metric decreases as the value of p increases.

Table 3.9 Separation measures and relative closeness of four alternatives under four metrics

Alternative	$p = 1$			$p = 1.5$			$p = 2$			$p = \infty$		
	S_i^+	S_i^-	C_i^*	S_i^+	S_i^-	C_i^*	S_i^+	S_i^-	C_i^*	S_i^+	S_i^-	C_i^*
A_1	0.1029	0.1515	0.5956	0.0665	0.1099	0.6231	0.0546	0.0984	0.6433	0.0385	0.0894	0.6991
A_2	0.1924	0.0620	0.2437	0.1380	0.0492	0.2630	0.1197	0.0439	0.2684	0.0894	0.0327	0.2677
A_3	0.0945	0.1599	0.6285	0.0677	0.1077	0.6140	0.0580	0.0920	0.6135	0.0447	0.0770	0.6325
A_4	0.1711	0.0834	0.3277	0.1163	0.0545	0.3190	0.1009	0.0458	0.3123	0.0894	0.0385	0.3009

3.4 Selection on PIS/NIS

Flexible selections on PIS/NIS in TOPSIS have sprung up in the literature due to many reasons. Opricovic and Tzeng (2004) recognized the problem on considering both PIS and NISand proposed the distance measure only to PIS in their VIKOR method. Chen et al. (2011) employed the concept of benchmark in management and proposed flexibility in setting the extremes or PIS/NIS. Kong (2011) identified the phenomenon of rank reversal in TOPSIS caused by its normalization procedure and suggested to consider the improved PIS/NIS to be determined by the DM according to his judgments or experience. Their values should be greater/less than the existing values in the decision matrix and to keep the normalized values between 0 and 1 to accommodate the most possible cases.

García-Cascales and Lamata (2012) studied the rank reversal phenomenon in the TOPSIS method and proposed two modifications on the concept of PIS and NISwith fictitious alternatives holding extreme values. These made the evaluation to be a kind of absolute mode and away from any sensitivity on ranking alternatives. Along the same stream, Cables et al. (2016) realized a similar problem in TOPSIS and presented a reference ideal method (RIM) that only deals with PIS and refers to extreme values. Mufazzal and Muzakkir (2018) introduced a proximity indexed value MCDM method, an adaptation of TOPSIS, and claimed that it minimizes rank reversals. In fact, their index values only count the Manhattandistance away from PIS. After examining the developments, we know that most works only prefer the reference point PIS, and this point could be chosen by DMs, who most likely select some larger or smaller values compared to the current performance values in the decision matrix. Nevertheless, the selection on the extremes is still a question.

3.5 Relative Closeness Formula

The decision rule of traditional TOPSIS is in line with a descending order of the values of relative closeness or ranking index $C_i = \frac{S_i^-}{S_i^+ + S_i^-}, i = 1,\ldots, m$, of alternatives in Eq. (2.6) (Hwang and Yoon 1981). The formula was designed to put the distance to NIS, S_i^-, as its numerator and the combined distance, S_i^+ and S_i^-, as its denominator. Its concept states that a compromise solution should be as close to PIS as possible and as far away from NISas possible. The rule is straightforward and reasonable in controlling the value within the interval [0, 1].

Lai et al. (1994) employed the concept of two distances forming a two-objective decision making or optimization problem and transformed it into a fuzzy compromise problem. Subsequently, Li (2009) presented a relative ratio method whose concept is similar to TOPSIS and VIKOR. His proposed the formula of relative closeness is:

$$C_i = \frac{S_i^-}{\max_i(S_i^-)} - \frac{S_i^+}{\min_i(S_i^+)}, \quad i = 1, \ldots, m, \text{ of alternatives} \tag{3.3}$$

where its PIS/NIS does not belong to the original decision matrix. Although there is no theoretical development, the index interval is $[-1, 1]$. Doukas et al. (2010) developed a linguistic TOPSIS to appraise renewable energy options. After converting linguistic variables into cardinal data, the TOPSIS steps could operate. Their relative closeness could be simply modified by

$$C_i = w^+ D_i^+ - w^- D_i^- \tag{3.4}$$

where $w^+ + w^- = 1$.

There is no reason provided for the recommendation, but the index is within the interval $[-1, 1]$.

Kuo (2017) investigated four different ranking indices or the possible combinations of relative closeness in TOPSIS. He observed that the weights given to Eq. (2.6) provide the same results of the traditional TOPSIS, shown by the following index:

$$C_i = \frac{w^- S_i^-}{w^+ S_i^+ + w^- S_i^-}, \ i = 1, \ldots, m, \ \text{of alternatives}, \tag{3.5}$$

where w^- and w^+ are the weights on two separation measure, respectively, and $w^- + w^+ = 1$. This means the relative closeness of Eq. (3.5) is independent of the weights on these two separation measures. He further compared the index $w^+ S_i^+ + w^- S_i^-$, Doukas et al.'s index (2010), and proposed the index below, Eq. (3.6):

$$C_i = w^+ \left(D_i^- / \sum_{i=1}^{m} D_i^- \right) - w^- \left(D_i^+ / \sum_{i=1}^{m} D_i^+ \right), \ i = 1, \ldots, m, \ \text{of alternatives}, \tag{3.6}$$

where the new index is within the interval $[-1, 1]$ instead of $[0, 1]$.

By deriving the formula, he proved that the necessary condition of an alternative to be the best is $C_i > \frac{w^+ - w^-}{m}$. After executing experiments on the consistency ratio, the suggested index is a better choice when the number of alternative has increased or there are different weights on the separation measures. Yoon and Kim (2017) carried out behavioral tendency through the concept of the S-shape value function, in which S_i^+ and S_i^- can be viewed as losses and gains individually. They proposed a value function, another ranking index, for the evaluation in Eq. (3.7):

$$C_i = S_i^- - \lambda S_i^+, i = 1, \ldots, m, \ \text{of alternatives}, \tag{3.7}$$

where parameter λ controls the steepness of the value function for losses; and $\lambda = 2.25$ as proposed by Tversky and Kahneman (1992).

It is obvious that the index is inside the interval $[-\lambda, 1]$.

Based on the terms appearing in Eqs. (3.3)–(3.7), the judgment goes according to their synthesized value in descending order, which is the same as the original formula does. Remark that we do not include the ranking indices by Ren et al. (2007) and Mufazzal and Muzakkir (2018) due to their proposals away from direct use of separation measures. To introduce a new formula is to incorporate a specific concentration, but it seems the development lacks any theoretical part and is only demonstrated by examples. Hence, a robust relative closeness formula or ranking index is still under development.

3.6 Ordinal Input

In the TOPSIS algorithm, there are two inputs, weights on criteria, and performance measures of all alternatives on criteria. The former is discussed in Sect. 3.2, including ordinal weight methods. Here, we shall figure out how to transform ordinal data into cardinal data on ordinal performance measures.

Coping with qualitative, linguistic, interval, or fuzzy variables, Hwang and Yoon (1981) recommended a 10-point scale, in Fig. 2.2, to transform the variable into cardinal variables for the benefit of TOPSIS calculation. The assignment is more subjective while its order could be kept. To give an example, the "high" on the benefit criterion is set to a scale 7. Why can it not be set to a scale 8? Another issue also arises. Two DMs' feelings might not exactly fit the same scale. In the mind of the first DM, the "high" means 7, while the "high" could be 8 inthe second one's mind. It is obvious that a more rigorous investigation is needed in the problem-solving process.

In the domain of fuzzy sets, Chen and Hwang (1992) summarized eight scales for the conversion. We replicate one of them, which covers five linguistic terms, containing very low, low, medium, high, and very high, to mimic the 5-point Likert scale shown in Fig. 3.3. A fuzzy set $\boldsymbol{A}$ in a universe $z \in \boldsymbol{U}$ is characterized by a membership function $\mu_A(z)$, which is associated with each element in $\boldsymbol{U}$ as a real number in the interval [0,1]. Observe in the figure that the membership functions of very low, low, medium, high, and very high are within the intervals [0, 0.2], [0.1, 0.4], [0.3, 0.7], [0.6, 0.9], and [0.8, 1.0], respectively. This conversion has an edge on its flexible range, which could take over vague information. However, the above-mentioned questions could not be resolved in this domain. Why not set high to the interval [0.65, 0.85]? If two DMs cooperate simultaneously to make a decision, then are both high intervals the same? Unfortunately, many fuzzy works perform their analyses by acting on their own discretion. For instance, Chen (2000), in his fuzzy group TOPSIS, offered fixed intervals for linguistic variables on weights of criteria and performance ratings and influenced many followers. A more reasonable approach is still needed.

As ordinal weights have been illustrated in the previous section, the ordinal data require only partial information and are reliable. The process of ordinal input on the performance measure on criteria will be examined here. The idea focuses on

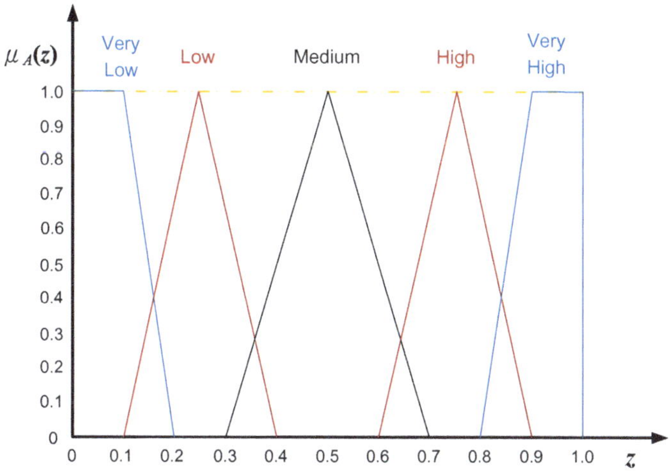

Fig. 3.3 Five linguistic terms conversion to fuzzy numbers (Chen and Hwang 1992)

the dominance of any alternative, through pair-wise comparisons fashioned on the performance measure of one alternative with the other alternatives', and sums up the dominances of this alternative with the others as accumulated dominances. Referring to the ordinal information operation in EVAMIX (Voogd 1983), we utilize a signum or sign function to count the dominance of the order comparison on each qualitative criterion for simplicity. The signum function that elicits the sign of a real number z is defined as (Wikipedia):

$$\operatorname{sgn}(z) = \begin{cases} 1, \text{ if } z > 0, \\ 0, \text{ if } z = 0, \\ -1, \text{ if } z < 0. \end{cases} \tag{3.8}$$

We next will handle a subset of the decision matrix of Eq. (2.1) in which all criteria are qualitative or ordinal as in Eq. (3.9).

$$\boldsymbol{D}_{qual} = \begin{matrix} \\ A_1 \\ A_2 \\ \vdots \\ A_i \\ \vdots \\ A_m \end{matrix} \begin{matrix} \begin{matrix} X_1 & X_2 & \cdots & X_j & \cdots & X_q \end{matrix} \\ \begin{bmatrix} x_{11} & x_{12} & \cdots & x_{1j} & \cdots & x_{1q} \\ x_{21} & x_{22} & \cdots & x_{2j} & \cdots & x_{2q} \\ \vdots & \vdots & \cdots & \vdots & \cdots & \vdots \\ x_{i1} & x_{i2} & \cdots & x_{ij} & \cdots & x_{iq} \\ \vdots & \vdots & \cdots & \vdots & \cdots & \vdots \\ x_{m1} & x_{m2} & \cdots & x_{mj} & \cdots & x_{mq} \end{bmatrix}_{m \times q} \end{matrix} \tag{3.9}$$

Here, A_i represents the alternative i, $i = 1, \ldots, m$; X_j denotes the criterion j, $j = 1, \ldots, q$; and the number of quantitative criteria p and the number of qualitative criteria q are restricted by $p + q = n$; their subscripts thus count from 1 to q.

As long as the ordinal information of performance measures on criterion j is by a partial order set, we can compare the orders of the performance measures. Here, the preference order of $x_{i1}, x_{i2}, \ldots, x_{iq}$ is donated by a function of their performances as $o(x_{i1})$, $o(x_{i2})$, …, and $o(x_{iq})$, respectively, for any alternative i, $i = 1, \ldots, m$. Notice that the orders of the performance measures are predefined in the partial order set, and they could be linguistic or semantic terms, or levels of achievement. Accordingly, a revised decision matrix is illustrated by the new notation.

$$
\boldsymbol{D}_{qual} = \begin{array}{c} \\ A_1 \\ A_2 \\ \vdots \\ A_i \\ \vdots \\ A_m \end{array} \begin{array}{c} \begin{array}{ccccccc} X_1 & X_2 & \cdots & X_j & \cdots & X_q \end{array} \\ \begin{bmatrix} o(x_{11}) & o(x_{12}) & \cdots & o(x_{1j}) & \cdots & o(x_{1q}) \\ o(x_{21}) & x_{22} & \cdots & x_{2j} & \cdots & x_{2q} \\ \vdots & \vdots & \cdots & \vdots & \cdots & \vdots \\ x_{i1} & x_{i2} & \cdots & x_{ij} & \cdots & x_{iq} \\ \vdots & \vdots & \cdots & \vdots & \cdots & \vdots \\ x_{m1} & x_{m2} & \cdots & x_{mj} & \cdots & o(x_{mq}) \end{bmatrix}_{m \times q} \end{array} \tag{3.10}
$$

A new defined signum function $s_j(o(x_{ij}), o(x_{i'j}))$, will give 1; if one is indifferent to the other, then this will give 0; if one is inferior to the other, then the function will give -1; after comparing two elements x_{ij} and $x_{i'j}$, i and $i' = 1, \ldots, m$ and $j = 1, \ldots, q$. The preference relation is thus formulated as follows (Shih 2016).

$$
s_j(o(x_{ij}), o(x_{i'j})) = \begin{cases} 1, \text{ if } o(x_{ij}) \succ o(x_{i'j}), \\ 0, \text{ if } o(x_{ij}) \approx o(x_{i'j}), \quad \forall i \text{ and } i', i \neq i' \\ -1, \text{ if } o(x_{ij}) \prec o(x_{i'j}). \end{cases} \tag{3.11}
$$

Based on the function, a return value is given after the comparison on any two elements at corresponding criterion j. Note that there is no need to distinguish the criteria of benefits and costs in the above expression since the comparison is based on the orders that are the preferences in DMs' minds.

The dominance measure $s_j(o(x_{ij}), o(x_{i'j}))$ is the comparison of the performance orders of any two alternatives A_i and $A_{i'}$, respectively, on criterion j. Thus, we illustrate a dominance matrix $\boldsymbol{S}_j$ by the following pair-wise comparisons on criterion j, $j = 1, \ldots, q$.

$$
\boldsymbol{S}_j = \begin{array}{c} A_1 \\ A_2 \\ \vdots \\ A_i \\ \vdots \\ A_m \end{array} \overset{\begin{array}{cccccc} A_1 & A_2 & \cdots & A_i & \cdots & A_m \end{array}}{\begin{bmatrix} - & s_j(o(x_{1j}), o(x_{2j})) & \cdots & s_j(o(x_{1j}), o(x_{ij})) & \cdots & s_j(o(x_{1j}), o(x_{mj})) \\ s_j(o(x_{2j}), o(x_{1j})) & - & \cdots & s_j(o(x_{2j}), o(x_{ij})) & \cdots & s_j(o(x_{2j}), o(x_{mj})) \\ \vdots & \vdots & \cdots & \vdots & \cdots & \vdots \\ s_j(o(x_{ij}), o(x_{1j})) & s_j(o(x_{ij}), o(x_{2j})) & \cdots & - & \cdots & s_j(o(x_{ij}), o(x_{mj})) \\ \vdots & \vdots & \cdots & \vdots & \cdots & \vdots \\ s_j(o(x_{mj}), o(x_{1j})) & s_j(o(x_{mj}), o(x_{2j})) & \cdots & s_j(o(x_{mj}), o(x_{ij})) & \cdots & - \end{bmatrix}}_{m \times m} \tag{3.12}
$$

The accumulated dominance measure t_{ij} is a summation of the comparison on the performance of alternative A_i to all other alternatives $A_{i'}$ on criterion j.

$$
t_{ij} = \sum_{i=1}^{m} s_j(o(x_{ij}), o(x_{i'j})), i \neq i', \forall j \tag{3.13}
$$

An accumulated dominance vector on criterion j, $j = 1,\ldots, q$, is thus represented as follows.

$$
\boldsymbol{T}_j = \begin{array}{c} A_1 \\ A_2 \\ \vdots \\ A_i \\ \vdots \\ A_m \end{array} \begin{bmatrix} t_{1j} \\ t_{2j} \\ \vdots \\ t_{ij} \\ \vdots \\ t_{mj} \end{bmatrix}_{m \times 1}, j = 1, \ldots, q \tag{3.14}
$$

Being connected with the elements in the above accumulated dominance vector, we next obtain an adjusted dominance index Q_{ij} with respect to each criterion j, corresponding to alternative i, by the following expression within their maximum and minimum values of the above vector (Shih 2016).

$$
Q_{ij} = \frac{t_{ij} - \min_{i=1}^{m}(t_{ij}) + \delta}{\max_{i=1}^{m}(t_{ij}) - \min_{i=1}^{m}(t_{ij}) + \delta}, i = 1, \ldots, m \text{ and } j = 1, \ldots, q \tag{3.15}
$$

Observe that we add a small value δ for normalization in order to avoid the zero performance for further computation. In the traditional TOPSIS, a small value δ is equal to 1 for the worst value of the benefit criteria (Hwang and Yoon 1981).

It is clear that Eq. (3.2) is similar to the form of linear normalization II – max-min, and it generates a positive value less than 1. Nevertheless, an extra normalized operation is available by linear or vector normalization as described in Sect. 3.1. After the process, the subset of qualitative criteria could merge into quantitative criteria and is coherent with the traditional TOPSIS procedure for further decision.

Example 3.4 Conversion of qualitative criteria in the fighter selection problem.

The original fighter selection problem keeps the two qualitative criteria, reliability (high-low) and maneuverability (high-low), in Example 3.2 (Hwang and Yoon 1981). Four alternatives are evaluated, and we will manage the data through Eqs. (3.11) and (3.13)–(3.15).

First, create two dominance matrices $\boldsymbol{S}_j$ for the criteria of the reliability and maneuverability by pair-wise comparisons. Tables 3.10 and 3.11 illustrate the two dominance matrices and their adjust dominance indices as follows.

By applying Eq. (3.15), the adjusted dominance indices corresponding to alternatives obtained are presented in Tables 3.12 and 3.13 for these two criteria. The normalized values by vector, linear, and with max-min type are listed for comparison. In our opinion, ordinal data refer relatively to each other, and it is better to hold full levels of intensity or scales on their performance of alternatives so that the adjustment or normalization does not have a tendency for any kind of data distortion. On account of the small value added in Eq. (3.15), our proposed formula has an advantage at avoiding the zero value, which means it is not worth mentioning the performance in the evaluation.

Table 3.10 Dominance information and the adjust dominance indices of reliability criterion

Alternative	A_1	A_2	A_3	A_4	Accumulated dominance
A_1	–	1	−1	0	0
A_2	−1	–	−1	−1	−3
A_3	1	1	–	1	3
A_4	0	1	−1	–	0

Table 3.11 Dominance information and the adjust dominance indices of maneuverability criterion

Alternative	A_1	A_2	A_3	A_4	Accumulated dominance
A_1	–	1	1	1	3
A_2	−1	–	−1	0	−2
A_3	−1	1	–	1	1
A_4	−1	0	−1	–	−2

Table 3.12 Adjusted dominance indices of alternatives on reliability criterion

Alternative	Adjusted dominance index	Vector normalization value	Linear normalization II value, max–min
A_1	0.5714	0.4811	0.5000
A_2	0.1429	0.2887	0.0000
A_3	1.0000	0.6736	1.0000
A_4	0.5714	0.4811	0.0000

Table 3.13 Adjusted dominance indices of alternatives on maneuverability criterion

Alternative	Adjusted dominance index	Vector normalization value	Linear normalization II value, max-min
A_1	1.0000	0.6708	1.0000
A_2	0.1667	0.3727	0.0000
A_3	0.6667	0.5217	0.5000
A_4	0.1667	0.1667	0.0000

In the beginning of this section, we have posted two questions on the input to the evaluation process. The proposed approach has an edge at solving these two questions. The first one relates to the fixed scale or level. Since the approach is on a personal basis, it is a self-comparison by DMs' preferences or attitudes. It does not matter whether one DM judges an alternative to have a high level, while the other DM thinks it should be a medium level. The second one involves the scale system. The approach allows one DM to use a 5-point scale and the other to consider a 7-point scale, and so on. The approach only needs the DMs to set their scale systems before the evaluation and make a self-comparison through the different comparisons. Therefore, the concept of Shih (2016), though the work focusing on group decision, is rather flexible for the input and appropriate for real-world decision making.

3.7 Weights on Separation Measures

In traditional TOPSIS, criteria weights are given to the normalized matrix to form a normalized decision matrix, Eq. (2.3) in Step 3 of the TOPSIS algorithm. We find its weighting process is not straightforward when comparing the function of weighted metrics (Steuer 1986). Thus, Shih et al. (2007) switched the weighting process to Step 5 in the calculation of the separation measures to fit the theme of the weighted formula. This shift makes the TOPSIS algorithm consistent with common MCDM algorithms. Kaliszewski and Podkopaev (2016) have demonstrated that the traditional TOPSIS weighting process would result in a non-weighted normalized decision matrix, and the original idea of working with data of the same magnitude is lost. Following this observations, we adopt the procedure of weighting the separation measures as in modified TOPSIS.

To show differences, Example 3.5 compares ranks on two weighting steps under four common metrics using the fighter selection problem.

Example 3.5 Comparison of ranks of two weighting steps in the fighter selection problem.

Following the work of Example 3.3 of the fighter selection problem (Hwang and Yoon 1981), we will examine the effects of weights on the normalized decision matrix (traditional TOPSIS) and on separation measures (modified TOPSIS) under the metrics $p = 1$, 1.5, 2, and ∞, respectively. The ranking results illustrate in Table 3.14. There are no major differences in their rankings. In this simple example, alternatives A_1 and A_3 are usually ranked higher than those of alternatives A_2 and A_4.

In addition to the fighter selection example, Shih (2016) also compared these two versions in the case study of a personnel recruitment problem involving fifteen candidates and two DMs. The results showed some rank differences in the group decision-making environment, but there seemed to be no significant differences in rankings, and sensitivity analysis using different decision power found that the selection of the first two options was stable. In later chapters, we will discuss the issue in group TOPSIS.

3.8 Dependent Criteria

Typically MCDM methods assume criteria to be independent so that an additive preference structure can be used to simplify the process. If relationships among criteria are significantly dependent, we need to manage them somehow. A couple of techniques have been developed to deal with dependent criteria in MCDM. Criteria dependencies are classified into three types by Gölcük and Baykasolu (2016): structural, where dominance and dependency relations arise from the structure of criteria that are required a priori; causal, where cause and effect relationships are the underlying reason for dependency; and preferential, where relative importance among criteria can vary depending on the level of attainment on one or more criteria. Criteria dependence in TOPSIS is recognized as structural dependency, and criteria weights are determined in the initial stages of the analysis.

Criteria dependencies have been found to be an issue in a number of TOPSIS applications. In vendor selection problems, price is likely to be positively correlated with a number of other important criteria, such as product quality, vendor technology, and vendor support level. Xu et al. (2018) analyzed a waste recycling situation, obtaining criteria interactions from linguistic assessments by managers and experts, yielding fuzzy relationships between economic, environmental, and social factors. These interactions had an impact on the ranking of innovation strategies. In the field of medical prosthetic material selection, Jahan and Edwards (2013) found that incorporating correlation of objective and subjective weightings was very important in two applications (amalgam tooth filling material selection; hip joint prosthesis material selection). While a number of other approaches have been proposed to deal with criteria dependence, the use of Mahalanobis distance has proven highly popular, despite additional computational complexity. In the content below, we will discuss the partial concept of analytic network process (ANP) in dealing with criteria dependence for TOPSIS. Mahalanobis distance applies to TOPSIS for managing the correlation among criteria. Some alternative methods are also reviewed.

Table 3.14 Ranks of two weighting schemes of four alternatives under four metrics

Alternative	$p = 1$		$p = 1.5$		$p = 2$		$p = \infty$	
	Traditional	Modified	Traditional	Modified	Traditional	Modified	Traditional	Modified
A_1	2	2	1	2	1	1	1	1
A_2	4	4	4	4	4	3	4	4
A_3	1	1	2	1	2	2	2	2
A_4	3	3	3	3	3	4	3	3

3.8.1 ANP Concept

AHP is a well-known MCDM method, and has been extensively adopted around the world in a wide variety of decision situations (Zyoud and Fuchs-Hanusch 2017). In AHP, criteria, alternatives, and other factors are assumed to be independent in the hierarchical structure, and this assumption may not always be applicable to real-world situations. Saaty (1996, 2005) tried to release this assumption and developed an extensive model called the analytic network process (ANP), which does not require independence among the above elements. Saaty presented ANP tools for solution of resulting models. AHP only handles comparisons of preference intensities on the elements at each level. However, ANP also measures the preference for dependence and feedback within any two elements and their clusters. The latter would generate an abundance of comparisons, especially as the number of elements increases. Regardless of the complex comparisons, Saaty and Takizawa (1986) forged a pruned process to treat dependence within a hierarchy before developing ANP. Their core concept is to separate the procedure into three parts. First, there is no dependence on AHP. Then a step models dependent relationships within pair of elements. Afterwards, the priorities from the above two steps are multiplied together to generate a combined priority, which is the final priority with dependence. This simple procedure could help MCDM methods manage dependent criteria.

Let's see the process through a simple example.

Example 3.6 A vendor selection problem (Shyur and Shih 2006).

A Taiwanese company was looking for a vendor for its key components. Seven criteria were identified for the evaluation: on-time delivery (X_1), product quality (X_2), price/cost (X_3), facility and technology (X_4), responsiveness to customer needs (X_5), professionalism of the salesperson (X_6), and quality of relationship with vendors (X_7) (Shyur and Shih 2006).

A DM first utilized the pair-wise comparison of AHP to determine the weights on the criteria without dependence, and the comparison and the corresponding priority vector/weight on DM's preference are shown in Table 3.15. Next, two dependent relationships among the criteria were modeled: (i) product quality X2 was influenced

Table 3.15 Pair-wise comparisons and their priorities of the seven criteria without dependence

	X_1	X_2	X_3	X_4	X_5	X_6	X_7	Priority/weight
X_1	1	2	4	5	4	6	3	0.338
X_2	1/2	1	3	4	4	5	3	0.251
X_3	1/4	1/3	1	3	4	3	2	0.140
X_4	1/5	1/4	1/3	1	1/2	1/4	1/5	0.036
X_5	1/4	1/4	1/4	2	1	5	1	0.083
X_6	1/6	1/5	1/3	4	1/5	1	2/5	0.048
X_7	1/3	1/3	1/2	5	1	5/2	1	0.103

Table 3.16 Pair-wise comparisons and their priorities of dependence related to criterion X_2

	X_2	X_4	Priority/weight
X_2	1	5	0.833
X_4	1/5	1	0.167

Table 3.17 Pair-wise comparisons and their priorities of dependence related to criterion X_3

	X_2	X_3	X_7	Priority/weight
X_2	1	1/2	11/10	0.231
X_3	2	1	5	0.609
X_7	10/11	1/5	1	0.160

by facility and technology, and (ii) price/cost X3 was influenced by product quality and the quality of the relationship with vendors. For instance, a couple of questions are raised, "Which criterion will influence X3 more: X2 or X7? And how much more?" The responses to these pair-wise comparisons of preference of two dependent relationships are illustrated in Tables 3.16 and 3.17. The rightmost column of each table is the corresponding priority vector or weight, respectively. The final weights of the criteria $\boldsymbol{Ws}$ could be obtained by synthesizing the previous two priorities as illustrated in Eq. (3.16). We can compare the results with and without dependent relationships to illustrate the influence of dependence.

$$W_s = \begin{bmatrix} w_{X_1} \\ w_{X_2} \\ w_{X_3} \\ w_{X_4} \\ w_{X_5} \\ w_{X_6} \\ w_{X_7} \end{bmatrix} = \begin{bmatrix} 1 & 0 & 0 & 0 & 0 & 0 & 0 \\ 0 & 0.833 & 0.231 & 0 & 0 & 0 & 0 \\ 0 & 0 & 0.609 & 0 & 0 & 0 & 0 \\ 0 & 0.167 & 0 & 1 & 0 & 0 & 0 \\ 0 & 0 & 0 & 0 & 1 & 0 & 0 \\ 0 & 0 & 0 & 0 & 0 & 1 & 0 \\ 0 & 0 & 0.160 & 0 & 0 & 0 & 1 \end{bmatrix} \begin{bmatrix} 0.338 \\ 0.251 \\ 0.140 \\ 0.036 \\ 0.083 \\ 0.048 \\ 0.103 \end{bmatrix} = \begin{bmatrix} 0.338 \\ 0.242 \\ 0.085 \\ 0.078 \\ 0.083 \\ 0.048 \\ 0.125 \end{bmatrix} \quad (3.16)$$

The difference between assuming independence in Eq. (3.16) and considering dependence in Table 3.15 is shown by weight changes. Weight for X_2 increased from 0.242 to 0.251, weight for X_3 increased from 0.085 to 0.140, weight for X_4 decreased from 0.078 to 0.036, and weight for X_7 decreased from 0.125 to 0.103. The other four weights were unchanged.

3.8.2 *Mahalanobis Distance*

Mahalanobis distance (1936) is a widely used technique to improve analytic performance in the presence of dependent criteria in TOPSIS (Kovács et al. 2005). The distance nonlinearly correlates the internal components of the evaluation matrix by its own covariance (correlation) matrix. Wang et al. (2018) appraised the external performance of Chinese energy regulations, finding that correlation between performance indices caused a significant difference in TOPSIS results. That study found that correlation between performance indices caused significant difference in appraisal results when applying TOPSIS to analysis of Chinese energy regulation impact. Sheikh et al. (2019) applied TOPSIS with Mahalanobis distance calculation to prioritize northeastern Iranian districts with respect to natural hazards such as floods and landslides, finding that strong correlations among criteria needed to be included to improve result accuracy. Liu et al. (2019) also applied Mahalanobis distance in their TOPSIS model for regional agricultural water and soil resources in Heilongjiang Province in China to more accurately identify resource development strategy effectiveness.

Vega et al. (2014) note that TOPSIS is based on the assumption of independent attributes. In Step 2 of the TOPSIS algorithm (Chap. 2), the decision matrix is normalized using the Euclidean norm (2nd power) and its element r_{ij}, $r_{ij} = \frac{x_{ij}}{\sqrt{\sum_{i=1}^{m} x_{ij}^2}}$, where $i = 1,\ldots, m$ and $j = 1,\ldots, n$. If attributes are correlated, this incorporates redundant information. Mahalanobis (1936) offered a measure between two random variables x and y with the same probability distribution and a non-singular variance–covariance matrix $\boldsymbol{S}$:

$$d_m = \sqrt{(x - y)^{\mathrm{T}} S^{-1} (x - y)} \tag{3.17}$$

where $\boldsymbol{S} = \frac{1}{n-1} X_c^{\mathrm{T}} X_c$.

$\boldsymbol{X}$ is the data matrix with m objects in rows and n columns. $\boldsymbol{X}_{\mathrm{C}}$ is the centered matrix, and $\boldsymbol{X}_{\mathrm{C}} = \boldsymbol{X}$ – arithmetic mean. The Mahalanobis distance is reduced to the Euclidean distance when the covariance matrix is the identity matrix. If the covariance matrix is diagonal, the resulting distance measure is known as a standardized Euclidean distance (Wikipedia):

$$d_m = \sqrt{\sum_{j=1}^{n} \frac{(x_j - y_j)^2}{s_j^2}} \tag{3.18}$$

where s_j is the standard deviation of the x_j and y_j, $j = 1, \ldots, n$, over the sample set.

When we consider Mahalanobis distance to measure the distance from any alternative A_i with n criteria, x_j, to its PIS or NIS, i.e., y_j, $j = 1, \ldots, n$ here. Though the calculation is a bit complex, Eqs. (3.17) or (3.18), could be directly merged into the traditional TOPSIS algorithm, i.e., relative closeness for the ith alternative $C_i^* = \frac{d_{m_i}^-}{d_{m_i}^+ + d_{m_i}^-}$, $i = 1,\ldots, m$. In addition, if there is a weight set given, Eq. (3.17) will be the following equation.

$$d_m = \sqrt{(x - y)^{\mathrm{T}} \Delta^T S^{-1} \Delta (x - y)} \tag{3.19}$$

where $\Delta = \text{diag}(\sqrt{w_1}, \ldots, \sqrt{w_n})$ represents the squared root of the weights $w_1, \ldots, w_n$ assigned to the n criteria, respectively.

Originally, Mahalanobis distance was used to measure the distance between a point and a distribution (Wikipedia). When we consider it as a measure of the distance on dependent criteria, the amount of data should be large enough to form a distribution. Otherwise, the process could be questionable. In the literature, for example, Wang et al. (2018) considered 17 years of data (as alternatives), Sheikh et al. (2019) compared 14 districts in the evaluation, and Liu et al. (2019) evaluated 15 farms. These data could not meet the minimum requirement of a large same size. Additionally, the majority of related works focus on processing data dependence, which is not at the heart of criteria dependence. We need to take a very cautious step to differentiate between both dependences. In the meantime, Antuchevičiene et al. (2010) pointed out that the covariance matrix could be calculated directly from initial data or from normalized data. But if there are not a lot more alternatives than criteria, statistically inaccurate estimates are possible. Consequently, they suggested using correlation matrices rather than covariance matrices because correlation matrices do not depend on absolute values of initial data. Correlation and covariance matrices are coincident if the standard deviation of the initial data is 1. Thus, it is proper to normalize the data matrix to ensure the standard deviations of initial data are equal to 1. The argument also could not directly deal with the issue of criteria dependence.

We will demonstrate an example using the Mahalanobis distance for the TOPSIS procedure.

Example 3.7 A natural hazard evaluation of the districts of Golestan Province (Sheikh et al. 2019).

This study aimed to prioritize the districts of Golestan Province, northeast of Iran, with highly susceptible areas to natural hazards. The evaluation included three cost indices: landslide susceptibility (X_1), water erosion (X_2), and flood inundation susceptibility (X_3), with equal weights. There were 16 districts (A_1 to A_{16}) involved. The corresponding decision information is shown in Table 3.18 (Sheikh et al. 2019).

Following the Mahalanobis distance for the TOPSIS algorithm, Eq. (3.19), the ranking results are displayed in Table 3.19 with the corresponding separation measures. The TOPSIS results are listed on the right hand side for reference. The first six ranks of alternatives are the same, and the maximum deviation of the rest is three. Because there is big difference, we take Spearman's rank correlation test, and the coefficient is 0.9297, which means they are closely related. Furthermore, for purposes of comparison, the results from the Mahalanobis distance with the correlation matrix are obtainable. There are slight differences in rankings for alternatives A_5 and A_{10}. Their Spearman's coefficient is 0.9956 inthe example, which means both matrices are highly correlated. We make a further note that alternative A_{13} is the ideal solution which is uncommon in TOPSIS evaluation.

Table 3.18 Decision table of the districts of Golestan Province (Sheikh et al. 2019)

	X_1	X_2	X_3
A_1	15.600	1.506	33.606
A_2	0	0	2.574
A_3	29.374	0.152	12.316
A_4	0	0	16.834
A_5	19.278	5.179	36.884
A_6	21.638	1.416	37.888
A_7	16.503	2.535	8.191
A_8	21.724	0.277	17.052
A_9	12.415	0	1.834
A_{10}	57.855	1.318	35.284
A_{11}	20.159	1.045	30.686
A_{12}	19.069	4.606	38.313
A_{13}	0	0	1.783
A_{14}	0	0	3.073

Table 3.19 The results of natural hazard evaluation of the districts of Golestan Province

	Mahalanobis distance				Euclidean distance (traditional TOPSIS)			
	d_m^+	d_m^-	C_i^*	Rank	S_i^+	S_i^-	C_i^*	Rank
A_1	1.9432	7.4720	0.7936	8	0.1462	0.2304	0.6118	10
A_2	0.0028	7.0030	0.9996	2	0.0029	0.3445	0.9917	2
A_3	1.3128	3.7231	0.7393	10	0.1236	0.2597	0.6775	7
A_4	1.0200	9.6404	0.9043	5	0.0548	0.3284	0.8571	5
A_5	3.1227	3.2555	0.5104	13	0.2656	0.1542	0.3673	14
A_6	2.5748	7.5868	0.7466	9	0.1683	0.2155	0.5615	11
A_7	1.6151	2.6640	0.6226	11	0.1283	0.2278	0.6397	8
A_8	0.7865	4.9908	0.8639	6	0.1037	0.2648	0.7186	6
A_9	0.3555	4.9527	0.9330	4	0.0496	0.3145	0.8638	4
A_{10}	4.7885	2.8756	0.3752	14	0.2672	0.1643	0.3807	13
A_{11}	1.6680	6.6512	0.7995	7	0.1397	0.2329	0.6251	9
A_{12}	2.6841	3.8110	0.5867	12	0.2484	0.1568	0.3871	12
A_{13}	0.0000	6.9103	1.0000	1	0.0000	0.3456	1.0000	1
A_{14}	0.0075	7.0643	0.9989	3	0.0047	0.3438	0.9865	3

3.8.3 Other Approaches

There have been quite a few efforts to improve TOPSIS performance in the presence of dependent criteria. In the AHP field, instead of ANP, Ramik and Perzina (2010)

proposed a method using geometric mean aggregation, allowing for fuzzy pairwise comparisons and feedback dependencies among criteria after demonstrating the impact on rank of criteria dependency. Yang et al. (2008) used integrated fuzzy MCDM, with sub-criteria relationships mapped through structural modeling. A fuzzy AHP was then used to obtain criterion weights, with a non-additive fuzzy integral providing fuzzy performance measures. Overall scores for alternatives were generated with fuzzy synthetic utilities. Gürbüz et al. (2012) combined ANP, Choquet integrals and the MACBETH MCDM method to deal with conjunctive or disjunctive behaviors between criteria.

Choquet integrals have been used to deal with criteria interactions in multiple criteria settings (Choquet 1953), and it commonly handles interaction criteria by fuzzy numbers. This creates computational complications (Lourenzutti et al. 2017). Gomes et al. (2013) also used Choquet integrals to incorporate criteria dependencies in the TODIM model. To treat criteria dependence, Yu et al. (2018) proposed a method using cosines to take into account the coordination level of attributes. Chen (2019b) identified some flaws in that original presentation and suggested improvements. This approach is intuitively attractive and applicable with Excel tools.

In some decision contexts, the importance of the criteria can vary conditionally. In supply chain management, vendor performance might include price and payment conditions. The relative importance of these might vary depending upon demand scarcity conditions. Choquet integrals are fuzzy measures that relax the requirement for additivity. This has the issue of an exponential growth in the number of parameters, leading to computational difficulties. Lourenzutti et al. (2017) dealt with criteria interaction with TOPSIS and TODIM (Gomes et al. 2013), as TOPSIS and TODIM are able to consider groups of DMs with varying opinions, heterogeneous information, criteria interaction, and fuzzy and dynamic environments. Meng et al. (2021) proposed the reverse Choquet integral to define the importance of the ordered elements. Meanwhile, Arcidiacono et al. (2021) modeled the performance of sovereign ratings on the economic, governmental, and financial performance of the 28 European countries. The criteria interactions were modeled to include negative interactions between GSP per capita and exports/GDP, positive interactions between interest/revenues and net debt/GDP, and positive interactions between current account receipts/GDP and current account balance/GEDP. Choquet integrals were used and the impact on ratings was reported.

Some experts use the technique of decision making trial and evaluation laboratory (DEMATEL) for handling dependent criteria for the TOPSIS algorithm, e.g., Baykasoğlu et al. (2013). The DEMATEL is a method for identifying cause-and-effect chain components in a complicated system. It deals with using a visual structural model to evaluate interdependent interactions among components and identify the key ones (Si et al. 2018). This cause-and-effect relationship would be difficult to manage given the criteria dependence of TOPSIS, where criteria are usually at the same level, and this kind of works could thus be unpersuasive.

3.9 Incremental Analysis

Incremental analysis (IA), or marginal analysis, is an essential tool for evaluating mutually exclusive alternatives in engineering economics. Compared to the present worth analysis with absolute values, the values of internal rate of return (IRR) and benefit-cost (B/C) ratio are relative values. The latter two could not be chosen solely on the basis of large values of alternatives performed. The right decision should further examine the relative advantages of the alternatives (Blank and Tarquin 2017). IA can be defined as the examination of the differences between two alternatives from the aspects of benefits and costs, or inputs and outputs. Accordingly, the incremental B/C ratio, or incremental IRR, of two neighboring alternatives is presented to examine the differences between them. By emphasizing alternatives, in an ascending order of costs, DMs are deciding whether or not differential costs, or input, are justified by differential benefits, or output, with (Newnan et al. 2013). The differential benefit-cost ratio $\Delta B/\Delta C$, a cut-off value, should be greater than 1 traditionally for the selection on higher cost-associated alternative; otherwise, the lower cost-associated alternative is kept. Please check Fig. 3.4 on illustrating IA on successive pair-wise analysis for four alternatives, i.e., *X*, *Y*, *Z*, and *W*, assuming that these four are in ascending order according to their initial costs spent. It is valuable for checking their ratios of differential benefit, ΔB, to differential cost, ΔC, so that the best alternative can finally be recognized. Seeing that MCDM typically takes advantage of ratio values in judgment, IA could supply a fair evaluation of the inputs and outputs of targeted alternatives (Shih 2008).

While IA is applied to MCDM problems, their criteria should be separated into two parts: benefit and cost criteria, or output-oriented and input-oriented criteria. The resources used or inputs are categorized as costs, and the performances exhibited are classified as benefits. After splitting the criteria into two parts, the performance on the corresponding criteria of the alternatives in the two parts could be integrated into two ratio scores individually. Following the streamline of IA, the differential cost score is then compared to the differential benefit scores of two alternatives, and the most preferred alternative will be acquired in the final pair-wise comparison. The edge of IA for MCDM is to ensure the differentiation of alternatives for a robust decision. For instance, among the set of alternatives *X* (with a performance score of 0.85), *Y* (with a score of 0.87), and *Z* (with a score of 0.86), alternative *Y* is

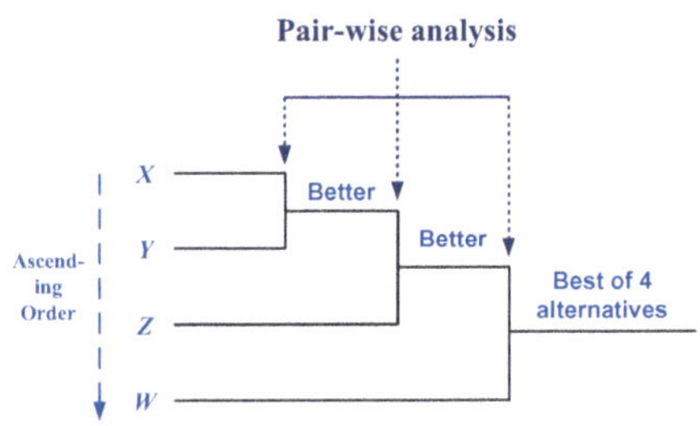

Fig. 3.4 Successive pair-wise analysis of four alternatives (Newnan et al. 2013)

chosen. DMs can believe that alternative Y is better than alternative Z, which is better than alternative X, and the judgment is made within a very narrow margin. DMs might have little confidence in the decision. Consequently, IA indeed delivers an efficient tool to guarantee the decision. Observe that the incremental B/C analysis and the incremental IRR analysis are illustrated in engineering economy. The latter seems not to make much sense for MCDM evaluation, and we keep the former as the framework for the evaluation.

The concept of IA could be exploited in most MCDM methods with two sub-problems: benefit-related and cost-related. After solving both sub-problems, two corresponding ranking indices are procured for processing their incremental B/C ratio, ΔB/ΔC here. Following the flow of Fig. 3.4, the final decision of the specified MCDM method is attained. For example, IA for AHP needs two hierarchies to measure benefit and cost priorities, individually, and combine them for the decision (Saaty 1980). Due to the fact that relative closeness is the ranking index of TOPSIS, the values of the relative closeness from the sub-problems, or for the benefit sub-problem, are employed in the procedure. In the latter case, the cost indices are acquired from the weighted normalized values, which depict the resources consumed. The proposed procedure is described as seven steps, combining with traditional TOPSIS, as follows (Shih 2008).

Step 1: Create a decision matrix ***D***, including benefit and cost criteria.

Gather decision information to establish the matrix. Please refer to Eq. (2.1).

Step 2: Construct the normalized decision matrix ***R***.

Please refer to Eq. (2.2).

Step 3: Construct the weighted normalized decision matrix ***V***.

IA needs to separate the decision information into benefit and cost criteria, two sub-problems, for later steps. The original weights assigned to all criteria, which summed to 1, should be modified so that all weights for benefit criteria and cost criteria should be kept to 1, respectively. Then the modified weights are multiplied to the corresponding normalized values from Step 2. Here we have to handle two sub-problems.

For the benefit criteria, go to Step 4. For cost criteria, take the summation of all weighted normalized values for each alternative as its cost index C_{ci}^*, and directly go to Step 7. A utility function $U(M)$ or an S-shaped value function (Kahneman and Tversky 1979) could also be used as the cost index. An exponential utility function suggested by Hillier and Lieberman (2014) is

$$U(M) = R\left(1 - e^{-\frac{M}{R}}\right) \tag{3.20}$$

where R denotes risk tolerance and M indicates monetary value.

Step 4: Determine the PIS and NIS for the benefit sub-problem.

There are two sets of PIS and NIS for two sub-problems.

Step 5: Calculate the separation measures between alternatives and their PIS/NIS for the benefit sub-problem.

Please consult Eqs. (2.4) and (2.5). There are m alternatives, and $2m$ separation measures are generated for the benefit sub-problem.

Step 6: Calculate the relative closeness/benefit index for the benefit sub-problem.

Please refer to Eq. (2.6). C_{bi}^* denote the relative closeness of the benefit criteria to alternative i, $i = 1, \ldots, m$, respectively. Thus, m numbers of relative closeness/benefit index for all alternatives are created for the benefit sub-problem.

Step 7: Execute incremental analysis.

(7.1) Rearrange all alternatives in ascending order by their cost indices.

(7.2) Select two adjacent alternatives with the least cost indices.

(7.3) Compute their incremental B/C ratio.

For example, alternatives X and Y are arranged by their cost indices in ascending order. Compare these two alternatives by their incremental B/C ratio, using the formula as follows.

$$\Delta B/\Delta C = \frac{C_{bY}^* - C_{bX}^*}{C_{cY}^* - C_{cX}^*} \tag{3.21}$$

(7.4) Make a choice on the value of $\Delta B/\Delta C$.

If $\Delta B/\Delta C \geq 1$ is true, alternative Y is the better choice, and it will be left for the next comparison. Otherwise, keep alternative X as the better one for the further process.

(7.5) Select another close alternative with it cost indices in ascending order for the pair-wise comparison.

Go to Step 7.3 to calculate their $\Delta B/\Delta C$ ratio for the judgment. If there is no more alternative to compare, the procedure stops. The left alternative at Step 7.4 would be the best alternative of all.

It should be noted that the cost-related indices could be accumulated from the performance of cost criteria of alternatives. In a broad sense, both cost-related and benefit-related indices could be transformed into utility indices if DM could provide the analogous utility functions (Shih 2008). Moreover, Kahneman and Tversky (1979) proposed the prospect theory to fit the behavioral decision making on gains and losses. Lee and Shih (2016) utilized their S-shaped function to modify cost-related indices for IA in extending the TODIM method. The cost part could be interpreted in different ways, e.g., utility functions for the cost index. The benefit index could also be established in the same fashion. It is worth noting that cost-related and benefit-related criteria should be completeness in formation, and the cost should imply the life cycle cost of the system.

Determining the cut-off value has drawn much attention. In engineering economics, since both benefits and costs are measured in money, it would be reasonable to choose 1 as the cut-off ratio. As the analysis was extended to MCDM, or intangible evaluation on benefits, some debates were posted as how to determine the equivalent value. Bernhard and Canada(1990) pointed out the problems with using benefit/cost ratio with AHP. Through the river-crossing example given by Saaty (1980), they found four different optimal solutions with cut-off value of less than 0.763, [0.763, 1.064], [1.064, 1.340], and greater than 1.340, respectively. Yang et al. (2004) criticized Bernhard and Canada's approach (1990) as incomplete and presented a simplified judgment procedure, but it still suffered from a debatable cut-off ratio. Since the normalized benefit and cost values in MCDM are no longer monetary measures, Shih (2008) tested utility functions, instead of weighted normalized values, to represent DMs' preferences. Consequently, the key to employing IA for MCDM is to adequately elicit utility functions of benefits and costs from DMs' aspect.

To show the process of IA for TOPSIS, we will illustrate an example from Jahanshahloo et al. (2009).

Example 3.8 A factory location problem (Jahanshahloo 2009).

Six cities, A_1 to A_6, in Iranwere assessed to find the best location for establishing a date factory. The cities were evaluated by four criteria, two of which related to input that is cost-oriented and the other two referred to output that is benefit-oriented. The former were distance from the border (km) X_1 and factory construction costs ($1000) X_2, while the latter were financing (percent) X_3 and regional production capacity (ton) X_4, and equal weights were considered. The first criterion was a real number, and the remaining three were in interval form (Jahanshahloo 2009). To demonstrate IA in a straightforward way, we take an average of the interval data on the latter three criteria and the modified decision information shown in Table 3.20.

To deal with IA, the problem should be separated into two sub-problems: benefits or output, and costs or input. The benefit sub-problem will be solved by the TOPSIS algorithm to acquire the relative closeness, or benefit index, of each alternative. The cost sub-problem will keep the essence of cost and thus obtain its weighted

Table 3.20 Decision table for the factory location problem

Alternative	Input		Output	
	X_1	X_2	X_3	X_4
A_1	1451	2834.5	45	170.0
A_2	843	4157.5	70	510.0
A_3	1125	3680.5	53	170.0
A_4	55	5898.5	80	386.5
A_5	356	4121.5	65	170.0
A_6	391	5426.5	80	431.0

Table 3.21 Relative closeness/benefit index of the factory location sub-problem

Alternative	S_i^+	S_i^-	C_i^*
A_1	0.2320	0.0000	0.0000
A_2	0.0306	0.2196	0.8778
A_3	0.2218	0.0245	0.0993
A_4	0.0748	0.1692	0.6935
A_5	0.2109	0.0611	0.2247
A_6	0.0478	0.1908	0.7996

normalized value as the cost index for each alternative. The benefit and cost indices are depicted in Tables 3.21 and 3.22, respectively. With the cut-off value equal to 1, the IA process is illustrated in Table 3.23 and the preferred alternative is A_2. The TOPSIS rankings are $A_4 \succ A_6 \succ A_2 \succ A_5 \succ A_3 \succ A_1$ for reference. Furthermore, after testing the cut-off value with 1.81, alternative A_4 would be preferred. It is apparent that the selected cut-off value is sensitive to the final choice. Furthermore, assume that an exponential utility function $U(M)$, Eq. (3.20), is elicited for factory construction costs X_2 by the DM. Here $R = 10{,}000$ is set for the attitude of risk aversion. To keep consistency with criterion X_1, the corresponding values from the equation are vector

Table 3.22 Composited weighted normalized vector/cost index of the factory location sub-problem

Alternative	Criterion		Cost index	Ascending order
	X_1	X_2		
A_1	0.3473	0.1293	0.4766	6
A_2	0.2018	0.1897	0.3915	4
A_3	0.2693	0.1679	0.4372	5
A_4	0.0132	0.2691	0.2823	2
A_5	0.0852	0.1880	0.2733	1
A_6	0.0936	0.2476	0.3412	3

Table 3.23 Results of IA of the factory location problem

Ascending order	Alternative	Cost index	benefit index	ΔB/ΔC ratio					Note
1	A_5	0.2733	0.2247						
2	A_4	0.2823	0.6935	51.87					
3	A_6	0.3412	0.7996		1.80				
4	A_2	0.3915	0.8778			1.56			The best one
5	A_3	0.4372	0.0993				−17.02		x
6	A_1	0.4766	0.0000					−10.31	x

Table 3.24 Results of IA of the factory location problem with utilities on construction costs

Ascending order	Alternative	Cost index	benefit index	$\Delta B/\Delta C$ ratio					Note
1	A_4	0.2688	0.6935						
2	A_5	0.2790	0.2247	−46.04					
3	A_6	0.3339	0.7996		1.63				
4	A_2	0.3969	0.8778			1.24			The best one
5	A_3	0.4459	0.0993				−15.89		X
6	A_1	0.4889	0.0000					−9.54	X

normalized and combined with the normalized values of X_1 as the cost index. The IA calculation and the results are shown in Table 3.24. Alternative A_2 is still preferred.

3.10 Summary and Discussions

This chapter has explored variants in normalization processes, weights on criteria, distance functions, selections of PIS/NIS, relative closeness formula, ordinal input, weights on separation measures, dependent criteria, and incremental analysis in TOPSIS developments. Eight examples are given to illustrate some of the characteristics on these variants. Regardless of these past fruitful works, there is still some theoretical support needed to yield a robust decision.

The TOPSIS algorithm is an integrated procedure. The nine issues mentioned above all impact the final preferred decision in different ways. These issues can arise in complex ways, especially in group contents. Multiple participants' contributions could be explored to provide an integrated group procedure for effective real world problem solving.

References

Acuña, C., Liern, V., Pérez-, B.: Normalization in TOPSIS-based approaches with data of different nature: application to the ranking of mathematical videos. Ann. Oper. Res. **296**, 541–569 (2021). https://doi.org/10.1007/s10479-018-2945-5

Alfares, H.K., Duffuaa, S.O.: Assigning cardinal weights in multi-criteria decision making based on ordinal ranking. J. Multi-Criteria Decis. Anal. **15**, 125–133 (2009). https://doi.org/10.1002/mcda.420

Antuchevičiene, J., Zavadskas, E.K., Zakarevičius, A.: Multiple criteria construction management decisions considering relations between criteria. Technol. Econ. Dev. Econ. **16**, 109–125 (2010). https://doi.org/10.3846/tede.2010.07

Arcidiacono, S.G., Corrente, S., Greco, S.: Robust stochastic sorting with interacting criteria hierarchically structured. Eur. J. Oper. Res. **292**, 735–754 (2021). https://doi.org/10.1016/j.ejor.2020.11.024

Barron, F.H., Barrett, B.E.: Decision quality using ranked attribute weights. Manage. Sci. **42**, 1515–1523 (1996). https://doi.org/10.1287/mnsc.42.11.1515

Baykasoğlu, A., Kaplanoğlu, V., Durmuşoğlu, Z.D.U., Şahin, D.: Integrating fuzzy DEMATEL and fuzzy hierarchical TOPSIS methods for truck selection. Expert Syst. Appl. **40**, 899–907 (2013). https://doi.org/10.1016/j.eswa.2012.05.046

Bernhard, R.H., Canada, J.R.: Some problems in using benefit/cost ratios with the analytic hierarchy process. Eng. Econ. **36**, 56–65 (1990). https://doi.org/10.1080/00137919008903031

Blank, L., Tarquin, A.: Engineering Economy, 8th edn. McGraw-Hill, New York (2017). ISBN: 9780073523439

Cables, E., Lamata, M.T., Verdegay, J.L.: RIM-reference ideal method in multicriteria decision making. Inf. Sci. **337–338**, 1–10 (2016). https://doi.org/10.1016/j.ins.2015.12.011

Chang, C.H., Lin, J.J., Lin, J.H., Chiang, M.C.: Domestic open-end equity mutual fund performance evaluation using extended TOPSIS method with different distance approaches. Expert Syst. Appl. **37**, 4642–4649 (2010). https://doi.org/10.1016/j.eswa.2009.12.044

Charnes, A., Cooper, W.W., Ferguson, R.O.: Optimal estimation of executive compensation by linear programming. Manage. Sci. **1**, 138–151 (1955)

Chen, C.T.: Extensions of the TOPSIS for group decision-making under fuzzy environment. Fuzzy Sets Syst. **114**, 1–9 (2000). https://doi.org/10.1016/S0165-0114(97)00377-1

Chen, P.: Effects of normalization on the entropy-based TOPSIS method. Expert Syst. Appl. **136**, 33–41 (2019a). https://doi.org/10.1016/j.eswa.2019.06.035

Chen, P.: A novel coordinated TOPSIS based on coefficient of variation. Mathematics **7**(614), 1–17 (2019b). https://doi.org/10.3390/math7070614

Chen, S.J., Hwang, C.L.: Fuzzy Multiple Attribute Decision Making: Methods and Applications. Springer, Berlin (1992). ISBN: 9783540549987

Chen, Y., Kilgour, D.M., Hipel, K.W.: An extreme-distance approach to multiple criteria ranking. Math. Comput. Model. **53**, 646–658 (2011). https://doi.org/10.1016/j.mcm.2010.10.001

Choo, E.W., Schoner, B., Wedley, W.C.: Interpretation of criteria weights in multicriteria decision making. Comput. Ind. Eng. **37**, 527–541 (1999). https://doi.org/10.1016/S0360-8352(00)00019-X

Choquet, G.: Theory of capacities. Annales De L'institut Fourier **5**, 131–295 (1953). https://doi.org/10.5802/aif.53

Chu, A.T.W., Kalaba, R.E., Spingarn, K.A.: A comparison of two methods for determining the weights of belonging to fuzzy sets. J. Optim. Theory Appl. **27**, 531–538 (1979). https://doi.org/10.1007/BF00933438

Cook, R.L., Stewart, T.R.: A comparison of seven methods for obtaining subjective descriptions of judgmental policy. Organ. Behav. Hum. Perform. **13**, 31–45 (1975). https://doi.org/10.1016/0030-5073(75)90004-5

Cook, W.D., Kress, M.: Deriving weights from pairwise comparison ratio matrices: an axiomatic approach. Eur. J. Oper. Res. **37**, 355–362 (1988). https://doi.org/10.1016/0377-2217(88)90198-1

Çelen, A.: Comparative analysis of normalization procedures in TOPSIS method: With an application to Turkish deposit banking market. Informatica **25**, 185–208 (2014). https://doi.org/10.15388/Informatica.2014.10

Dawes, R.M., Corrigan, B.: Linear models in decision making. Psychol. Bull. **81**, 91–106 (1974). https://doi.org/10.1037/h0037613

Diakoulaki, D., Martotas, G., Papayannakis, L.: Determining objective weights in multiple criteria problems: the CRITIC method. Comput. Oper. Res. **22**, 763–770 (1995). https://doi.org/10.1016/0305-0548(94)00059-H

Doukas, H., Karakosta, C., Psarras, J.: Computing with words to assess the sustainability of renewable energy options. Expert Syst. Appl. **37**, 5491–5497 (2010). https://doi.org/10.1016/j.eswa.2010.02.061

Eckenrode, R.T.: Weighting multiple criteria. Manage. Sci. **12**, 180–192 (1965). https://doi.org/10.1287/mnsc.12.3.180

Fan, Z.P., Zhang, X., Chen, F.D., Liu, Y.: Multiple attribute decision making considering aspiration-levels: a method based on prospect theory. Comput. Ind. Eng. **65**, 341–350 (2013). https://doi.org/10.1016/j.cie.2013.02.013

García, F., Guijarro, F., Moya, I.: A goal programming approach to estimating performance weights for ranking firms. Comput. Oper. Res. **37**, 1597–1609 (2010). https://doi.org/10.1016/j.cor.2009.11.018

García-Cascales, M.S., Lamata, M.T.: On rank reversal and TOPSIS method. Math. Comput. Model. **56**, 123–132 (2012). https://doi.org/10.1016/j.mcm.2011.12.022

Gomes, L.F.A.M., Lima, M.M.P.P.: TODIM: basics and application to multi-criteria ranking of projects with environmental impacts. Found. Comput. Decis. Sci. **16**, 113–127 (1991)

Gomes, L.F.A.M., Lima, M.M.P.P.: From modelling individual preferences to multi-criteria ranking of discrete alternatives: a look at prospect theory and the additive difference model. Found. Comput. Decis. Sci. **17**, 171–184 (1992)

Gomes, L.F.A.M., Machado, M.A.S., Rangel, L.A.D.: Behavioral multi-criteria decision analysis: the TODIM method with criteria interactions. Ann. Oper. Res. **211**, 531–548 (2013). https://doi.org/10.1007/s10479-013-1345-0

Gölcüka, İ, Baykasoğlua, A.: Analysis of DEMATEL approaches for criteria interaction handling within ANP. Expert Syst. Appl. **46**, 346–366 (2016). https://doi.org/10.1016/j.eswa.2015.10.041

Gürbüz, T., Alptekin, S.E., Alptekin, G.I.: A hybrid MCDM methodology for ERP selection problem with interacting criteria. Decis. Support Syst. **54**, 206–214 (2012). https://doi.org/10.1016/j.dss.2012.05.006

Hamming, R.W.: Error detecting and error correcting codes. Bell Syst. Tech. J. **29**, 147–160 (1950). https://doi.org/10.1002/j.1538-7305

Hellinger, E.: Neue begründung der theorie quadratischer formen von unendlichvielen veränderlichen. Journal Für Die Reine Und Angewandte Mathematik (in German) **136**, 210–271 (1909). https://doi.org/10.1515/crll.1909.136.210

Hillier, F., Lieberman, G.: Introduction to Operations Research, 10th edn. McGraw-Hill, New York (2014). ISBN: 9781259162985

Hwang, C.L., Yoon, K.: Multiple Attribute Decision Making: A State-of-the-Art Survey. Springer-Verlag, Berlin (1981). ISBN: 9783540105589

Izadikhah, M.: Using the Hamming distance to extend TOPSIS in a fuzzy environment. J. Comput. Appl. Math. **231**, 200–207 (2009). https://doi.org/10.1016/j.cam.2009.02.102

Jahan, A., Edwards, K.L.: Weighting of dependent and target-based criteria for optimal decision-making in materials selection process: biomedical applications. Mater. Des. **49**, 1000–1008 (2013). https://doi.org/10.1016/j.matdes.2013.02.064

Jahan, A., Edwards, K.L.: A state-of-the-art survey on the influence of normalization techniques in ranking: improving the materials selection process in engineering design. Mater. Des. **65**, 335–342 (2015). https://doi.org/10.1016/j.matdes.2014.09.022

Jahan, A., Mustapha, F., Sapuan, S.M., Ismail, M.Y., Bahraminasab, M.: A framework for weighting of criteria in ranking stage of material selection process. Int. J. Adv. Manuf. Technol. **58**, 411–420 (2012). https://doi.org/10.1007/s00170-011-3366-7

Jahanshahloo, G.R., Lotfi, F.H., Davoodi, A.R.: Extension of TOPSIS for decision-making problems with interval data: interval efficiency. Math. Comput. Model. **49**, 1137–1142 (2009). https://doi.org/10.1016/j.mcm.2008.07.009

Jee, D.H., Kang, K.J.: A method for optimal material selection aided with decision making theory. Mater. Des. **21**, 199–206 (2000). https://doi.org/10.1016/S0261-3069(99)00066-7

Jones, D.F., Mardle, S.J.: A distance-metric methodology for the derivation of weights from a pairwise comparison matrix. J. Oper. Res. Soc. **55**, 869–875 (2004). https://doi.org/10.1057/palgrave.jors.2601745

Kahneman, D., Tversky, A.: Prospect theory: an analysis of decision under risk. Econometrica **47**, 263–292 (1979). https://doi.org/10.2307/1914185

Kaliszewski, I.: Rankings Revisited. Decyzje **2018**(30), 75–84 (2018). https://doi.org/10.7206/DEC.1733-0092.115

Kaliszewski, I., Podkopaev, D.: Simple additive weighting—a metamodel for multiple criteria decision analysis methods. Expert Syst. Appl. **54**, 155–161 (2016). https://doi.org/10.1016/j.eswa.2016.01.042

Kirkwood, C.W., Corner, J.L.: The effectiveness of partial information about attribute weights for ranking alternatives in multiattribute decision making. Organ. Behav. Hum. Decis. Process. **54**, 456–476 (1993). https://doi.org/10.1006/obhd.1993.1019

Kong, F.: Rank reversal and rank preservation in TOPSIS. Adv. Mater. Res. **204–210**, 36–41 (2011). https://doi.org/10.4028/www.scientific.net/amr.204-210.36

Kovács, P., Petres, T., Tóth, L.: A new measure of multicollinearity in linear regression models. Int. Stat. Rev. **73**, 405–412 (2005). https://doi.org/10.1111/j.1751-5823.2005.tb00156.x

Kuo, T.: A modified TOPSIS with a different ranking index. Eur. J. Oper. Res. **260**, 152–160 (2017). https://doi.org/10.1016/j.ejor.2016.11.052

Lai, Y.J.: TOPSIS for MODM. Eur. J. Oper. Res. **76**, 486–500 (1994). https://doi.org/10.1016/0377-2217(94)90282-8

Larichev, O.I.: Cognitive validity in design of decision-aiding techniques. J. Multi-Criteria Decis. Anal. **1**, 127–138 (1992). https://doi.org/10.1002/mcda.4020010303

Lee, Y.S., Shih, H.S.: Incremental analysis for generalized TODIM. CEJOR **24**, 901–922 (2016). https://doi.org/10.1007/s10100-015-0427-2

Li, D.F.: Relative ratio method for multiple attribute decision making problems. Int. J. Inf. Technol. Decis. Mak. **8**, 289–311 (2009). https://doi.org/10.1142/S0219622009003405

Liu, D., Qi, X., Fu, Q., Li, M., Zhu, W., Zhang, L., Faiz, M.A., Khan, M.I., Li, T., Cui, S.: A resilience evaluation method for a combined regional agricultural water and soil resource system based on weighted Mahalanobis distance and a Gray-TOPSIS model. J. Clean. Prod. **229**, 667–679 (2019). https://doi.org/10.1016/j.jclepro.2019.04.406

Lourenzutti, R., Krohling, R.A.: The Hellinger distance in multicriteria decision making: an illustration to the TOPSIS and TODIM methods. Expert Syst. Appl. **41**, 4414–4421 (2014). https://doi.org/10.1016/j.eswa.2014.01.015

Lourenzutti, R., Krohling, R.A., Reformat, M.Z.: Choquet based TOPSIS and TODIM for dynamic and heterogeneous decision making with criteria interaction. Inf. Sci. **408**, 41–69 (2017). https://doi.org/10.1016/j.ins.2017.04.037

Ma, J., Fan, Z.P., Huang, L.H.: A subjective and objective integrated approach to determine attribute weights. Eur. J. Oper. Res. **112**, 397–404 (1999). https://doi.org/10.1016/S0377-2217(98)00141-6

Mahalanobis, P.C.: On the generalised distance in statistics. Proc. Natl. Inst. Sci. India **2**, 49–55 (1936). https://doi.org/10.1007/s13171-019-00164-5(Reprinted)

Meng, F., Chen, S.M., Tang, J.: Multicriteria decision making based on bi-direction Choquet integrals. Inf. Sci. **555**, 339–356 (2021). https://doi.org/10.1016/j.ins.2020.10.055

Mufazzal, S., Muzakkir, S.M.: A new MCDM method based on proximity indexed value for minimizing rank reversals. Comput. Ind. Eng. **119**, 427–438 (2018). https://doi.org/10.1016/j.cie.2018.03.045

Newnan, D.G., Lavelle, J.P., Eschenbach, T.G.: Engineering Economic Analysis, 12th edn. Oxford University Press, New York (2013). ISBN: 9780199339273

Nijkamp, P.: Stochastic quantitative and qualitative multicriiteria analysis for environmental design. Reg. Sci. **39**, 175–199 (1977a). https://doi.org/10.1111/j.1435-5597.1977.tb01006.x

Nijkamp, P. A. van Delft: Multicriteria Analysis and Regional Decision Making. Springer, Leiden (1977b). ISBN: 9789020706895

Olson, D.L.: Comparison of weights in TOPSIS models. Math. Comput. Model. **40**, 721–727 (2004). https://doi.org/10.1016/j.mcm.2004.10.003

Olson, D.L., Dorai, V.K.: Implementation of the centroid method of Solymosi and Dombi. Eur. J. Oper. Res. **60**, 117–129 (1992). https://doi.org/10.1016/0377-2217(92)90339-B

Opricovic, S., Tzeng, G.H.: Compromise solution by MCDM methods: a comparative analysis of VIKOR and TOPSIS. Eur. J. Oper. Res. **156**, 445–455 (2004). https://doi.org/10.1016/S0377-2217(03)00020-1

Pavličić, D.M.: Normalisation effects the results of MADM methods. Yugoslav J. Oper. Res. **11**, 251–265 (2001)

Peldschus, F.: Recent findings from numerical analysis in multi-criteria decision making. Technol. Econ. Dev. Econ. **24**, 1695–1717 (2018). https://doi.org/10.3846/20294913.2017.1356761

Ramík, J., Perzina, R.: A method for solving fuzzy multicriteria decision problems with dependent criteria. Fuzzy Optim. Decis. Making **9**, 123–141 (2010). https://doi.org/10.1007/s10700-010-9078-x

Ren, L., Zhang, Y., Wang, Y., Sun, Z.: Comparative analysis of a novel M-TOPSIS method and TOPSIS. Appl. Math. Res. eXpress Article ID abm005, 1–10 (2007). https://doi.org/10.1093/amrx/abm005

Rezaei, J.: Best-worst multi-criteria decision-making method. Omega **53**, 49–57 (2015). https://doi.org/10.1016/j.omega.2014.11.009

Riabacke, M., Danielson, M., Ekenberg, L.: State-of-the-art prescriptive criteria weight elicitation. Adv. Decis. Sci. **2012**, Article ID 276584, 1–24 (2012). https://doi.org/10.1155/2012/276584

Saaty, T.L.: The Analytic Hierarchy Process. McGraw-Hill, New York (1980). ISBN: 9780070543713

Saaty, T.L.: Decision Making with Dependence and Feedback: The Analytic Network Process. RWS Publications, Pittsburgh, Pennsylvania (1996). ISBN: 9780962031793

Saaty, T.L.: Theory and Applications of the Analytic Network Process: Decision Making with Benefits, Opportunities, Costs and Risks. RWS Publications, Pittsburgh, Pennsylvania (2005). ISBN: 9781888603064

Saaty, T.L., Takizawa, M.: Dependence and independence—from linear hierarchies to nonlinear networks. Eur. J. Oper. Res. **26**, 229–237 (1986). https://doi.org/10.1016/0377-2217(86)90184-0

Schmitt, N.: Comparison of subjective and objective weighting strategies in changing task situations. Organ. Behav. Hum. Perform. **21**, 171–188 (1978). https://doi.org/10.1016/0030-5073(78)90048-X

Schoemaker, P.J.H., Waid, C.C.: An experimental comparison of different approaches to determining weights in additive utility models. Manage. Sci. **28**, 182–195 (1982). https://doi.org/10.1287/mnsc.28.2.182

Sheikh, V., Kornejady, A., Ownegh, M.: Application of the coupled TOPSIS–Mahalanobis distance for multi-hazard-based management of the target districts of the Golestan Province. Iran. Nat. Haz. **96**, 1335–1365 (2019). https://doi.org/10.1007/s11069-019-03617-0

Shih, H.S.: An incremental analysis for MCDM with an application to group TOPSIS. Eur. J. Oper. Res. **186**, 720–734 (2008). https://doi.org/10.1016/j.ejor.2007.02.012

Shih, H.S.: A mixed-data evaluation in group TOPSIS with differentiated decision power. Group Decis. Negot. **25**, 537–565 (2016). https://doi.org/10.1007/s10726-015-9451-9

Shih, H.S., Shyur, H.J., Lee, E.S.: An extension of TOPSIS for group decision making. Math. Comput. Model. **45**, 801–813 (2007). https://doi.org/10.1016/j.mcm.2006.03.023
Shyur, H.-J., Shih, H.-S.: A hybrid MCDM model for strategic vendor selection. Math. Comput. Model. **44**, 749–761 (2006). https://doi.org/10.1016/j.mcm.2005.04.018
Si, S.L., You, X.Y., Liu, H.C., Zhang, P.: DEMATEL technique: a systematic review of the state-of-the-art literature on methodologies and applications. Math. Probl. Eng. **2018**(3696457), 1–33 (2018). https://doi.org/10.1155/2018/3696457
Steuer, R.E.: Multiple Criteria Optimization: Theory, Computation, and Application. John Wiley, New York (1986). ISBN: 9780471888468
Stillwell, W.G., Seaver, D.A., Edwards, W.: A comparison of weight approximation techniques in multiattribute utility decision making. Organ. Behav. Hum. Perform. **28**, 62–77 (1981). https://doi.org/10.1016/0030-5073(81)90015-5
Triantaphyllou, E.: Multi-Criteria Decision Making Methods: A Comparative Study. Kluwer, Dordrecht (2000). ISBN: 9781475731576
Tversky, A., Kahneman, D.: Advances in prospect theory: cumulative representation of uncertainty. J. Risk Uncertain. **5**, 297–323 (1992). https://doi.org/10.1007/BF00122574
Tzeng, G.-H., Chen, T.-Y., Wang, J.-C.: A weight-assessing method with habitual domains. Eur. J. Oper. Res. **110**, 342–367 (1998). https://doi.org/10.1016/S0377-2217(97)00246-4
Vega, A., Aguarónb, J., García-Alcaraza, J., Moreno-Jiménez, J.M.: Notes on Dependent Attributes in TOPSIS. Procedia Computer Science **31**, 308–317 (2014). https://doi.org/10.1016/j.procs.2014.05.273
Voogd, H.: Multicriteria Evaluation for Urban and Regional Planning. Pion, London (1983). ISBN: 9780850861068
Wang, Z.X., Li, D.D., Zheng, H.H.: The external performance appraisal of Chinaenergy regulation: an empirical study using a TOPSIS method based on entropy weight and Mahalanobis distance. Int. J. Environ. Res. Public Health **15**(236), 1–18 (2018). https://doi.org/10.3390/ijerph15020236
Weber, M., Borcherding, K.: Behavioral influences on weight judgments in multiattribute decision making. Eur. J. Oper. Res. **67**, 1–12 (1993). https://doi.org/10.1016/0377-2217(93)90318-H
Wiecek, M.M., Ehrgott, M., Engau, A.: Countinuous multiobjective programming. In: Greco, S., Ehrgott, M., Figueira, J.R. (Eds.) Multiple Criteria Decision Analysis: State of the Art Surveys, pp. 739–815, Chap. 18, 2nd Edn. Springer, New York (2016). ISBN 9781493930937
Xu, Y., Zhang, L., Yeh, C.H., Liu, Y.: Evaluating WEEE recycling innovation strategies with interacting sustainability-related criteria. J. Clean. Prod. **190**, 618–629 (2018). https://doi.org/10.1016/j.jclepro.2018.04.078
Yager, R.R.: On ordered weighted averaging aggregation operators in multicriteria decisionmaking. IEEE Trans. Syst. Man Cybern. **18**, 183–190 (1988). https://doi.org/10.1016/10.1109/21.87068
Yager, R.R.: Quantifier guided aggregation using OWA operators. Int. J. Intell. Syst. **11**, 49–73 (1996). https://doi.org/10.1002/(SICI)1098-111X(199601)11:1%3c49::AID-INT3%3e3.0.CO;2-Z
Yang, J.L., Chiu, H.N., Tzeng, G.H., Yeh, R.H.: Vendor selection by integrated fuzzy MCDM techniques with independent and interdependent relationships. Inf. Sci. **178**, 4166–4183 (2008). https://doi.org/10.1016/j.ins.2008.06.003
Yang, K.L., Chu, P., Chouhuang, W.T.: Note on incremental benefit/cost ratios in analytic hierarchy process. Math. Comput. Modell. **39**, 279–286 (2004). https://doi.org/10.1016/S0895-7177(04)90011-8
Yin, L., Shyur, H.J.: A robust group multiple attributes decision-making method based on risk preference of the decision makers. Int. J. Appl. Sci. Eng. **15**, 33–46 (2018). https://doi.org/10.6703/IJASE.201802_15(1).033
Yoon, K.: A reconciliation among discrete compromise solutions. J. Oper. Res. Soc. **38**, 277–286 (1987). https://doi.org/10.1057/jors.1987.44
Yoon, K.P., Hwang, C.-L.: Multiple Attribute Decision Making: An Introduction. Sage Publications, Thousand Oaks, CA (1995). ISBN: 9780803954861

Yoon, K.P., Kim, W.K.: The behavioral TOPSIS. Expert Syst. Appl. **89**, 266–272 (2017). https://doi.org/10.1016/j.eswa.2017.07.045

Yu, L., Yang, W., Duan, Y., Long, X.: A study on the application of coordinated TOPSIS in evaluation of robotics academic journals. Math. Probl. Eng. **2018**(5456064), 1–7 (2018). https://doi.org/10.1155/2018/5456064

Yu, P.L.: A class of solutions for group decision problems. Manage. Sci. **19**, 936–946 (1973). https://doi.org/10.1287/mnsc.19.8.936

Zanakis, S.H., Solomon, A., Wishart, N., Dublish, S.: Multi-attribute decision making: a simulation comparison of selection methods. Eur. J. Oper. Res. **107**, 507–529 (1998). https://doi.org/10.1016/S0377-2217(97)00147-1

Zardari, N.H., Ahmed, K., Shirazi, S.M., Yusop, Z.B.: Weighting Methods and their Effects on Multi-Criteria Decision Making Model Outcomes in Water Resources Management. Springer, Cham (2015). ISBN: 9783319125855

Zavadskas, E.K., Turskis, Z.: A new logarithmic normalization method in games theory. Informatica **19**, 303–314 (2008). https://doi.org/10.15388/Informatica.2008.215

Zeleny, M.: A concept of compromise solutions and the method of the displaced ideal. Comput. Oper. Res. **1**, 479–496 (1974). https://doi.org/10.1016/0305-0548(74)90064-1

Zeng, Q.L., Li, D.D., Yang, Y.B.: VIKOR method with enhanced accuracy for multiple criteria decision making in healthcare management. J. Med. Syst. **37**(9908), 1–9 (2013). https://doi.org/10.1007/s10916-012-9908-1

Zyoud, S.H., Fuchs-, D.: A bibliometric-based survey on AHP and TOPSIS techniques. Expert Syst. Appl. **78**, 158–181 (2017). https://doi.org/10.1016/j.eswa.2017.02.016

Chapter 4
Group Aspects of TOPSIS

Group decision making (GDM) or collective decision-making is a situation faced when individuals collectively make a choice from the alternatives before them. GDM is one of the most successful models for generating support from other stakeholders, building consensus, and encouraging creativity (Wikipedia). This chapter will discuss the background, processes, consensus, and applications of group decision, especially on the developments of TOPSIS (Technique for Order Preference by Similarity to Ideal Solution).

4.1 Backgrounds of Group Decisions

A group is defined as a number of people, interacting and interdependent, that are put together or considered as a unit. Decisions by a group are rather common in our society, from a family decision to buy a car, a firm decision to recruit some sales engineers, a government decision to select a nuclear waste site, to United Nations decision to set sustainable goals. The techniques of MCDM (Multi-Criteria Decision Making) makes decisions under conflicting criteria, and often are applied in group contexts which have the added factor of group preferences (Hwang and Lin 1987).

There is some debate on group decisions versus individual decisions. The advantages of GDM include the following (Hitt et al. 2015): Groups can gather more knowledge and information, contributing to better options; (ii) groups frequently exhibit superior judgment from diverse perspectives when evaluating alternatives in a complex environment; (iii) group participation in decision-making generates a higher level of acceptance and satisfaction; and (iv) group decision-making can bring about growth of the group members. On the other hand, its disadvantages include (Robbins and Judge 2019): (i) group decisions are more time-consuming to make decisions than individuals; (ii) conformity pressures force group members to squash any overt disagreement; (iii) groups are sometimes dominated by one or

H.-S. Shih and D. Olson, *TOPSIS and its Extensions: A Distance-Based MCDM Approach*, Studies in Systems, Decision and Control 447,
https://doi.org/10.1007/978-3-031-09577-1_4

a few members, who may reduce the group's overall effectiveness; and (iv) group decisions suffer from ambiguous responsibility. Decision makers need to consider these factors in seeking better decisions.

Compared to a more general concept of the term "group", a team can be defined as a number of people who do something together as a group, but the team involves generating positive synergy through coordinated effort (Robbins and Judge 2019). A team is a cluster of individuals working toward specific goals or outcomes, and it is rather common in organizations as production teams, service teams, management teams, and project teams (Hitt et al. 2015). From a decision-making viewpoint, we could roughly differentiate both by their outputs. A team decision presumes only one output. For instance, in the determination of weights on criteria in MCDM, its output would be one set, although the determination could be implicitly supplied by the group members; at the least, there exists some kind of consent. If group members have different weight sets which can be aggregated into one set from individual decisions, it is a group decision (Shih et al. 2007). Hence, GDM explicitly presents the effort of group members. In this book, we consider groups instead of teams.

4.2 Group Decision Processes

A group's members first seek to understand the background of the problem and then obtain relevant decision information. The group process then involves aggregation, collecting individuals' preferences into a group preference. The aggregation of individuals' preferences could be considered to involve three stages, i.e., input, process, and output. Dias and Clímaco (2000) classified the participation of individuals at the input and output stages. At the input stage we the input data required could include a criteria weight set or thresholds on criteria. These could be obtained by the Delphitechnique prior to the MCDM core process (Shih et al. 2007). Analysis at the output stage follows the MCDM process, and could involve a social choice function (i.e., the Borda function) to aggregate individual ranks. We denote process stage analysis as aggregation within the MCDM process. Shih (2008) defined two types of external aggregation: pre-operation (aggregating criteria ratings and their weights) and post-operation (aggregating individuals' ranks), which match Dias and Clímaco (2000). Table 4.1 classifies the aggregation stages for some common MCDM methods and provides the corresponding examples. Observe that the compromise or agreement on the aggregation could be reached through feedback and discussion among group members (Bezerra et al. 2014).

The above table gives us a general concept of group decision making to support MCDM via three stages. In the same vein, we list some typical TOPSIS group support in Table 4.2 and show the detailed process in Fig. 4.1.The work in the upper part involves group decision at the input stage, the light green block in the figure. The performance measure and the weight under each criterion would be the information for TOPSIS for aggregation, as in the work of Parkan and Wu (1998). At the input stage, there are approaches in addition to those shown in Fig. 4.1. Conversely, if each

Table 4.1 Aggregation stages in common MCDM by group decisions (Shih 2021)

Aggregation stage	Comparison of the classifications		Example	Note
	Dias and Clímaco (2000)	Shih (2008)		
Input stage	Transfer the inputs of group members into the group's input by an operator	External aggregation with pre-operation	(1) Aggregation of individual judgments for AHP (Forman and Peniwati 1998) (2) Weighted sum of subjective ratings on some criteria by experts for TOPSIS (Parkan and Wu 1998)	Aggregation at this stage could be considered as teamwork instead of group decision
Process stage	Not available	Internal aggregation	(1) Aggregation of individual priorities for AHP (Forman and Peniwati 1998) (2) Establishment of group separation measure for TOPSIS (Shih et al. 2007) (3) Weighting individual net flows to be the global evaluation for PROMETHEE (Macharis et al. 1998)	This stage of aggregation is rather diverse, depending on the procedures of MCDM techniques
Output stage	Transfer the results of group members into the group's result by an operator	External aggregation via post-operation	(1) Aggregation of individual ranks from TOPSIS into a group rank by Borda function (Shih et al. 2004) (2) Accumulation of the ranks of multi-judges from PROMETHEE through social welfare functions (Colson 2000)	Aggregation at this stage operates by social choice functions for the group decision

Note AHP: Analytic Hierarchy Process; PROMETHEE: Preference Ranking Organization METHod for Enrichment of Evaluations

member would like to make his or her own decision using the TOPSIS procedure, and sum up the rank by applying the Borda function, it is at the output stage, as in the work of Shih et al. (2001), as illustrated by the dot-dash brown color lines in the figure. There are other options involved for the output stages which are listed in the lower leftmost part of Fig. 4.1.

It is worth mentioning aggregation at the process stage, shown in the light blue block at the center of Fig. 4.1. For TOPSIS, Shih (2016) and Shih et al. (2007) have defined the group separation measure as an operation of individuals' separation measures. Then the group's relative closeness is obtained for the final decision.

Table 4.2 Some typical group supports for TOPSIS

Aggregation stage	Support content	Illustration	Note
Input stage	Performance measures	Weighted sum of performance on subjective criteria (1–9 scale)	Parkan and Wu (1998)
	Fuzzy performance measures and fuzzy criteria weights	Arithmetical mean of two fuzzy measures	Chen (2000)
	Fuzzy performance measures on subjective criteria and fuzzy criteria weights	Arithmetical mean of two fuzzy measures	Chu (2002)
Process stage	Separation measures	Arithmetical and geometric means of two separation measures	Shih et al. (2007)
		Arithmetical and geometric means of two separation measures with differentiated decision powers of individuals	Shih (2016)
	Differences of relative closeness (preferential differences) and rank (preferential priorities)	Integration of preferential differences and priorities for the group TOPSIS procedure at the second stage	Huang and Li (2012)
Output stage	Ranks	Borda function of individuals' ranks	Shih et al. (2001, 2004)
		Borda function of individuals' ranks under different criteria sets	Pochampally and Gupta (2004)

Suppose that there are K members or DMs in a group. Decision matrix $\boldsymbol{D}^k$, $k = 1, \ldots, K$, represents the kth member, as follows.

$$\boldsymbol{D}^k = \begin{array}{c} \\ A_1 \\ A_2 \\ \vdots \\ A_i \\ \vdots \\ A_m \end{array} \begin{array}{c} \begin{array}{cccccc} X_1 & X_2 & \ldots & X_j & \ldots & X_n \end{array} \\ \begin{bmatrix} x_{11}^k & x_{12}^k & \ldots & x_{1j}^k & \ldots & x_{1n}^k \\ x_{21}^k & x_{22}^k & \ldots & x_{2j}^k & \ldots & x_{2n}^k \\ \vdots & \vdots & \ldots & \vdots & \ldots & \vdots \\ x_{i1}^k & x_{i2}^k & \ldots & x_{ij}^k & \ldots & x_{in}^k \\ \vdots & \vdots & \ldots & \vdots & \ldots & \vdots \\ x_{m1}^k & x_{m2}^k & \ldots & x_{mj}^k & \ldots & x_{mn}^k \end{bmatrix}_{m \times n} \end{array} \tag{4.1}$$

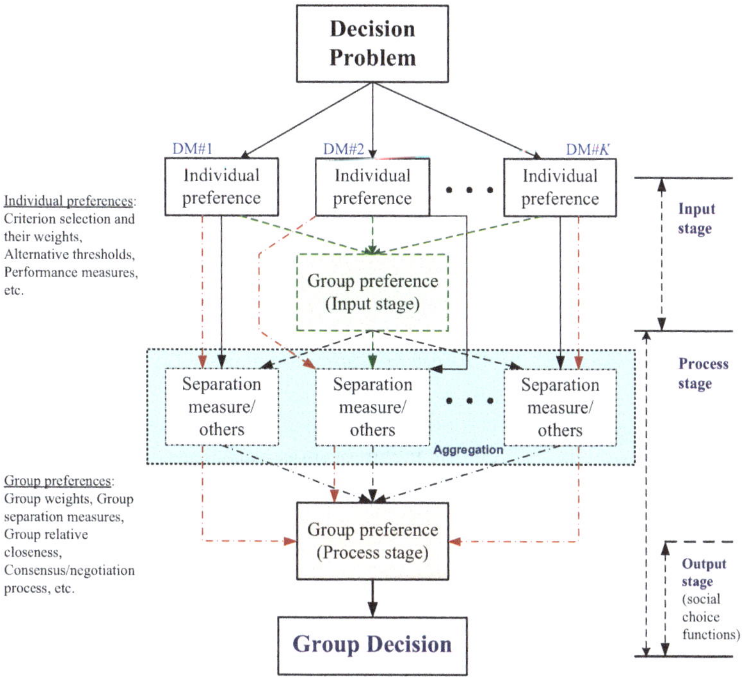

Fig. 4.1 Group support for TOPSIS

where A_i represents the alternative i, $i = 1, \ldots, m$; X_j donates the criterion j, $j = 1, \ldots, n$. The element of $\boldsymbol{D}^k$, x_{ij}^k, indicates the performance rating of alternative A_i w.r.t. attribute X_j by member k, $k = 1, \ldots, K$. Observe that there should be K decision matrices for K members of the group.

Following the TOPSIS algorithm in Sect. 2.2, each member will normalize his/her decision matrix as Eq. (4.1), weight the normalized matrix, determine PIS/NIS, and calculate the separation measures for PIS/NIS. Now there are K separation measures, represented by S_i^{K+} and S_i^{K-}, $k = 1, \ldots, K$, to PIS and NIS, respectively, for alternative i, $i = 1, \ldots, m$. Shih et al. (2007) proposed group measures, $\overline{S_i^+}$ and $\overline{S_i^-}$, as aggregates of the separation measures of all members, S_i^{K+} and S_i^{K-}, $k = 1, \ldots, K$, respectively, using geometric or arithmetical means of members' measures.

$$\overline{S_i^+} = S_i^{1+} \oplus \ldots \oplus S_i^{K+}, \text{ for alternative } i, \text{ and} \tag{4.2}$$

$$\overline{S_i^-} = S_i^{1-} \oplus \ldots \oplus S_i^{K-}, \text{ for alternative } i. \tag{4.3}$$

Here $\oplus$ serves as an operator, and geometric or arithmetical means are the most common operators in aggregation.

Additionally, the above aggregation could consider differential group member decision power (Tsui et al., 2015). A straightforward modification is to consider decision power as another weight attached to the corresponding member's separation measures (Baucells and Sarin 2003). Hence, Eqs. (4.2) and (4.3) will be revised subsequently as Eqs. (4.4) and (4.5).

$$\overline{S_i^+} = (u^1 \otimes S_i^{1+}) \oplus \ldots \oplus (u^K \otimes S_i^{1+}), \text{ for alternative } i, \text{ and} \tag{4.4}$$

$$\overline{S_i^-} = (u^1 \otimes S_i^{1+}) \oplus \ldots \oplus (u^K \otimes S_i^{K+}), \text{ for alternative } i. \tag{4.5}$$

Here $\otimes$ is generally a multiplication operator, and u^k, $k = 1, \ldots, K$, is the decision power of member k, and $\sum_{k=1}^{K} u^k = 1$.

There are a couple of options for the operation at the process stage, e.g., weighted arithmetic mean, weighted geometric mean, or other variations. Taking the weighted arithmetic mean of all individuals' separation measures, we get the following group separation measures from PIS and NIS:

$$\overline{S_i^+} = \left(\sum_{k=1}^{K} u^k S_i^{k+}\right) \Big/ K, \text{ for alternative } i, \text{ and} \tag{4.6}$$

$$\overline{S_i^-} = \left(\sum_{k=1}^{K} u^k S_i^{k-}\right) \Big/ K, \text{ for alternative } i, \tag{4.7}$$

where $i = 1, \ldots, m$; $k = 1, \ldots, K$.

When using the weighted geometric mean, the group separation measures from PIS and NIS are as follows.

$$\overline{S_i^+} = \prod_{k=1}^{K} (S_i^{k+})^{u^k}, \text{ for alternative } i, \text{ and} \tag{4.8}$$

$$\overline{S_i^-} = \prod_{k=1}^{K} (S_i^{k-})^{u^k}, \text{ for alternative } i. \tag{4.9}$$

In the following, the final ranking is determined by the group's relative closeness as the group preference. The formulation of the ith alternative A_i can be represented as:

$$\overline{C_i^*} = \frac{\overline{S_i^-}}{\overline{S_i^+} + \overline{S_i^-}}, \quad i = 1, \ldots, m, \tag{4.10}$$

where $0 \leq \overline{C_i^*} \leq 1$. It is obvious that the higher the index value, the better the alternative's performance will be.

Note that Eqs. (4.6), (4.7), and (4.10) are extensions of the TOPSIS algorithm of Sect. 2.2, preserving the theme of distance-based concepts. The extra work would not generate much computational burden, and the process efficiency of TOPSIS for the group could be maintained.

Example 4.1 A manager's recruitment and selection problem (Shih et al. 2007).

A chemical company was recruiting an online manager. The human resources department of the company provided data on relevant selection tests, which all were categorized as benefit criteria. These seven included knowledge tests (language test, professional test and safety rule test), skill tests (professional skills and computer skills), and interviews (panel interview and 1-on-1 interviews), i.e., $X_1, \ldots, X_7$. There were 17 qualified candidates, donated by A to Q, on the list, and the decision was the responsibility of four DMs as a group. The evaluation data under objective (the first five) and subjective criteria (the latter two) are listed in Tables 4.3 and 4.4. In addition, the weights of criteria elicited by DMs are presented in Table 4.5 as follows (Shih et al. 2007).

Table 4.3 Decision matrix I – objective criteria for 17 candidates

Candidate	Knowledge tests			Skill tests	
	Language test (X_1)	Profession test (X_2)	Safety rule test (X_3)	Professional skill (X_4)	Computer skill (X_5)
A	80	70	87	77	76
B	85	65	76	80	75
C	78	90	72	80	85
D	75	84	69	85	65
E	84	67	60	75	85
F	85	78	82	81	79
G	77	83	74	70	71
H	78	82	72	80	78
I	85	90	80	88	90
J	89	75	79	67	77
K	65	55	68	62	70
L	70	64	65	65	60
M	95	80	70	75	70
N	70	80	79	80	85
O	60	78	87	70	66
P	92	85	88	90	85
Q	86	87	80	70	72

Table 4.4 Decision matrix II – subjective criteria of by four DMs for 17 candidates

Candidate	DM#1		DM#2		DM#3		DM#4	
	Panel interview (X_6)	1-on-1 interview (X_7)	Panel interview (X_6)	1-on-1 interview (X_7)	Panel interview (X_6)	1-on-1 interview (X_7)	Panel interview (X_6)	1-on-1 interview (X_7)
A	80	75	85	80	75	70	90	85
B	65	75	60	70	70	77	60	70
C	90	85	80	85	80	90	90	95
D	65	70	55	60	68	72	62	72
E	75	80	75	80	50	55	70	75
F	80	80	75	85	77	82	75	75
G	65	70	70	60	65	72	67	75
H	70	60	75	65	75	67	82	85
I	80	85	95	85	90	85	90	92
J	70	75	75	80	68	78	65	70
K	50	60	62	65	60	65	65	70
L	60	65	65	75	50	60	45	50
M	75	75	80	80	65	75	70	75
N	80	70	75	72	80	70	75	75
O	70	65	75	70	65	70	60	65
P	90	95	92	90	85	80	88	90
Q	80	85	70	75	75	80	70	75

Table 4.5 Weight sets on six criteria by four DMs

Decision maker	Criteria						
	X_1	X_2	X_3	X_4	X_5	X_6	X_7
DM#1	0.066	0.196	0.066	0.130	0.130	0.216	0.196
DM#2	0.042	0.112	0.082	0.176	0.118	0.215	0.255
DM#3	0.060	0.134	0.051	0.167	0.100	0.203	0.285
DM#4	0.047	0.109	0.037	0.133	0.081	0.267	0.326

For further reference, Table 4.6 displays the results of the relative closeness and the alternatives rankings of the four DMs by traditional TOPSIS (vector normalization on original data, and Euclidean distance for measuring alternatives away from their PIS/NIS). There is a great difference in rank among the DMs. However, all their Spearman rank correction coefficients are above the value of 0.7034, which is greater than the critical value 0.414 associated with the significant level $\alpha = 0.05$.

Table 4.7 illustrates the results of Example 4.1 with arithmetical and geometric means on the individuals' separation measures using group TOPSIS. For comparative purposes, the table collects the results from three stages for aggregation. At the input stage, the decision information, performance ratings and weights, of all DMs are

Table 4.6 Results of the relative closeness and ranking of the four DMs

Candidate	DM#1		DM#2		DM#3		DM#4	
	C_i^*	Rank	C_i^*	Rank	C_i^*	Rank	C_i^*	Rank
A	0.5700	7	0.6548	4	0.5108	10	0.7947	4
B	0.4089	15	0.3328	13	0.5528	7	0.4078	15
C	0.8178	2	0.7082	3	0.8170	2	0.9216	2
D	0.4716	11	0.3011	16	0.5190	9	0.4652	12
E	0.5439	9	0.5407	7	0.2168	16	0.5478	10
F	0.6717	5	0.6422	5	0.7137	4	0.6075	7
G	0.4517	13	0.3264	14	0.4587	13	0.5251	11
H	0.4681	12	0.4491	11	0.5093	11	0.7809	5
I	0.7951	3	0.8943	2	0.8960	1	0.9444	1
J	0.4944	10	0.5266	8	0.5372	8	0.4454	13
K	0.0927	17	0.1689	17	0.2395	15	0.4149	14
L	0.2101	16	0.3130	15	0.1280	17	0.0545	17
M	0.5665	8	0.5972	6	0.5082	12	0.5561	9
N	0.6050	6	0.5258	9	0.5812	6	0.6076	6
O	0.4263	14	0.4296	12	0.4129	14	0.3445	16
P	0.9219	1	0.9268	1	0.7961	3	0.9094	3
Q	0.6956	4	0.4616	10	0.6397	5	0.5563	8

lumping together as a unit for the later TOPSIS procedure. At the output stage, all DMs independently obtain ranks which are aggregated using the Borda function. Note that there is no value of relative closeness at this stage. At the process stage, following the Steps 1 to 5 of the traditional TOPSIS algorithm (Chap. 2), each DM constructs his or her normalized decision matrix, determines PIS/NIS, and calculates the separation measures. Then the group's separation measures would be obtained by the aggregation of individuals', and the group's relative closeness $\overline{C_i^*}$ could be calculated, via Eqs. (4.4), (4.5), and (4.10), for the group ranks.

Observe that the rank differences from these three stages are closer than that of individuals, which means the group decision reduces the rank gaps under some kind of compromise. The first stage data are tied ranked by the Borda function, but rank aggregation does not provide an accurate decision. Furthermore, the ranking gaps in the group decision converge, as shown in Table 4.7. All Spearman rank correction coefficients are above the value of 0.9547 among each pair of comparisons, which is greater than the value of 0.7034 among the DMs from Table 4.6 given the significant level $\alpha = 0.05$. This could be the edge of a group decision. In addition, the same case is analyzed using the modified TOPSIS procedure, weighting on separation measures. Table 4.8 presents the corresponding results compared to Table 4.7. The rankings in Table 4.8 are rather consistent, and all Spearman coefficients are greater than the value of 0.9951. This indicates a benefit of weighting on separation measures.

Table 4.7 Ranking results of 17 candidates by traditional group TOPSIS

Candidate	Arithmetical mean				Geometric mean				Output stage
	Input stage		Process stage		Input stage		Process stage		
	$\overline{C_i^*}$	Rank	$\overline{C_i^*}$	Rank	$\overline{C_i^*}$	Rank	$\overline{C_i^*}$	Rank	Borda Rank
A	0.6306	5	0.8526	5	0.6345	5	0.9164	6	4
B	0.3819	14	0.8375	12	0.3871	14	0.8968	11	13
C	0.8401	3	0.8691	3	0.8416	3	0.9460	3	1
D	0.3979	12	0.8350	14	0.3973	12	0.8909	15	14
E	0.4141	11	0.8375	13	0.4044	11	0.8961	12	9
F	0.6643	4	0.8540	4	0.6685	4	0.9294	4	5
G	0.3937	13	0.8378	11	0.3955	13	0.8924	14	12
H	0.5214	9	0.8450	9	0.5197	9	0.8993	10	11
I	0.9308	2	0.8726	2	0.9326	2	0.9622	2	1
J	0.4660	10	0.8426	10	0.4705	10	0.9068	9	10
K	0.1266	16	0.8245	16	0.1330	16	0.8687	17	16
L	0.1032	17	0.8187	17	0.1006	17	0.8724	16	17
M	0.5337	8	0.8470	8	0.5372	8	0.9125	8	8
N	0.5729	7	0.8470	7	0.5781	7	0.9127	7	7
O	0.3573	15	0.8348	15	0.3595	15	0.8937	13	15
P	0.9352	1	0.8727	1	0.9366	1	0.9679	1	1
Q	0.5807	6	0.8490	6	0.5829	6	0.9182	5	6

4.3 Consensus in Group TOPSIS

In the preceding sections, it is assumed that decision information is provided in advance by a team or task group; at the very least, there is some kind of agreement on the use of TOPSIS and associated algorithms, which may involve laborious negotiations. In reality, this assumption could be too optimistic, and we need to deal with the topic of consensus and how to reach it. The advantages of group decision making with a high level of consensus can lead to a final decision that is more accepted and appreciated (Rodríguez et al. 2018). There are three major issues involved: (i) how to measure the consensus; (ii) how to define a poor consensus; and (iii) how to improve the level of consensus in case of poor consensus. There are a great number of papers involving these issues (Dong et al. 2018), especially in fuzzy domain (Szmidt and Kacprzyk 2003).

Table 4.8 Ranking results of 17 candidates by modified group TOPSIS

Candidate	Arithmetical mean				Geometric mean				Output stage
	Input stage		Process stage		Input stage		Process stage		
	$\overline{C_i^*}$	Rank	$\overline{C_i^*}$	Rank	$\overline{C_i^*}$	Rank	$\overline{C_i^*}$	Rank	Borda Rank
A	0.6257	5	0.6262	5	0.6278	5	0.6232	5	5
B	0.4324	13	0.4589	14	0.4314	13	0.4597	14	14
C	0.6817	3	0.6552	4	0.6823	3	0.6558	4	4
D	0.4248	14	0.4743	13	0.4242	14	0.4755	13	12
E	0.5748	6	0.5793	8	0.5771	6	0.5807	8	8
F	0.6624	4	0.6654	3	0.6633	4	0.6679	3	3
G	0.4390	12	0.4784	12	0.4388	12	0.4772	12	12
H	0.4929	11	0.5738	9	0.4886	11	0.5781	9	9
I	0.7817	2	0.8113	2	0.7788	2	0.8281	2	2
J	0.4983	10	0.5151	11	0.4982	10	0.5162	11	11
K	0.0210	17	0.0942	17	0.0281	17	0.0000	17	16
L	0.1645	16	0.1962	16	0.1565	16	0.1814	16	16
M	0.5329	8	0.5492	10	0.5327	8	0.5484	10	10
N	0.5139	9	0.5849	6	0.5155	9	0.5845	7	7
O	0.3645	15	0.4168	15	0.3648	15	0.4174	15	15
P	0.9547	1	0.9269	1	0.9557	1	0.9269	1	1
Q	0.5710	7	0.5833	7	0.5706	7	0.5846	6	6

4.3.1 Consensus Measures and Reaching

Ben-Arieh and Easton (2007) organized the four classes to measure the level of consensus relative to the group opinion: (i) based on a count of the number of experts who share the opinion with the group, (ii) based on the rank order of the alternatives by the group and by the experts, (iii) based on the distance between the experts and the group decision, and (iv) based on a similarity or dissimilarity measure between the experts and the group. The first class deals with counting of votes; similar to plurality elections, selecting the option that has more votes than any other among all candidates (Hwang and Lin 1987). The second class counts rank orders so that the preferred order or priority ranking of alternatives by each DM is included in the process. The Borda function and the Condorcet function are examples of this class (Hwang and Lin 1987). The third class examines the consensus among a set of preferences by a distance function. Minkowski metrics and rank-based distances are this class's illustrations (Cook 2006). A similarity measure is used quantify the closeness of experts' preferences. Although similarity has a close relationship with distance function, the fourth class measures consensus degrees and proximity by a similarity function, or the similarity of a pair of vectors in multi-dimensional (criteria) space. These comprise the Manhattandistance, Euclidean distance, and the sine and cosine of the angle between two vectors (Bryson 1996). The fuzzy matching function also provides countless contributions in this area, e.g., Chen (1988). It seems that the recent developments focus on the last two classes.

4.3.2 Consensus Reaching Process

In order to improve the level of consensus, individual DMs need to modify their opinions, seeking results more similar or close to the group consensus level in a consensus reaching process (CRP). This is considered an iterative and dynamic process, leading to a final group consensus-based solution after multiple rounds of discussions and modifications (Dong et al. 2018). Labella et al. (2018) divided the CRP scheme for a group into four phases: (i) gather preferences of group members, (ii) identify a consensus measure on collective preferences or between group members, (iii) consensus control by a minimum value of acceptable agreement, and (iv) consensus progress for increasing the level of agreement through feedback or automatic adjustment. Whereas they produced a general view of CRP, some details will be proposed in a multi-criteria environment.

In a group decision, the third dimension, DMs, is appended to the decision matrix in addition to the dimensions of alternatives and criteria. Our concentration could be expanded as the consensus on criteria and their weights, and alternatives with the whole evaluation in the phase of consensus measure. The former checks the deviations of the performance on any criterion or criteria weights by the group of members/individuals; the latter examines the discrepancy of total performance of an alternative, with multiple criteria, by the group's members. The former is known as consensus at the criterion level, whereas the latter is known as consensus at the alternative level. The data from both can offer profound insights into CRP. It is noted that only subjective criteria are counted at the criterion level as long as there is no difference in the evaluation of objective criteria. During the phase of consensus progress, there are two ways to update the group's preference. One is to improve the consensus level by an automatic process with system's suggestion. The other is to highlight deviations away from the mean of the group. Information could include specific members' values for relevant criteria and alternatives. The proposed improved values are also counted in some systems. Furthermore, because the agreement might not be reached, a bound on the number of runs or consumed time would be predetermined to make the process practicable and another option would be to not allow changes of preferences as well. The three ways described above would enable stopping. A new scheme of CRP is submitted in Fig. 4.2, which could provide a guideline to check the flow and the essential elements of CRP. Ma et al. (2014) suggested a three-level similarity measuring method of participant opinions by: (i) assessment level for the similarity of any criterion, (ii) criterion level for each criterion's similarity to the criteria set, and (iii) problem level for the similarity between two opinions by participants. It seems that the second level could not provide effective information for decision support. In addition, if scrutinizing the targets for consensus compared to the contents of

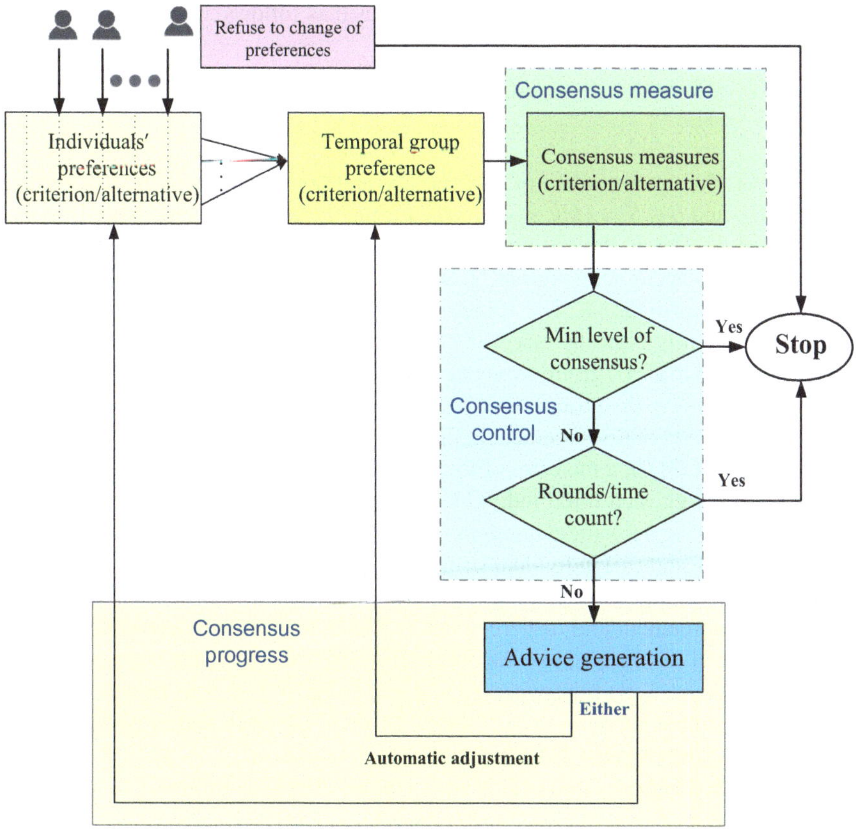

Fig. 4.2 Proposed consensus reaching process

Fig. 4.1, we can find various options, such as the number of criteria, the number of alternatives, weights on criteria, etc., outside Ma et al.'s scope. Single or multiple target selections will depend on the DMs' concentration. In the following sections, we will introduce two general techniques for CRP in Sects. 4.3.3 and 4.3.4.

4.3.3 Ngwenyama et al.'s Approach

Ngwenyama et al. (1996) presented an approach to support the facilitator in building consensus in GDM for AHP preference vectors. Based on the concept of similarity measure, they assessed individual preferences and offered information about the level of group consensus, subgroup coalescence, and areas of strong disagreement. Through use of the cosine function, any two preference vectors u^p and u^q of DMs p and q, respectively, in a multiple-dimensional space could be measured their degree of similarity. The agreement index (AGI) is defined as follows.

$$\mathrm{AGI}^{p,q} = \frac{u^p u^q}{|u^p||u^q|} \tag{4.11}$$

where $u^p u^q$ is the inner product of two vectors, and $|u^p|$ is the norm of vector u^p so as the $|u^q|$. p and $q \in S = \{k|k = 1,\ldots, K\}$ for the group with K members.

Note that $\mathrm{AGI}^{p,q}$ is the cosine similarity (Wikipedia). The value of $\mathrm{AGI}^{p,q}$ is close to 1 if two vectors are similar or coincide; the value reaches 0 if two vectors are dissimilar or orthogonal. The value of $\mathrm{AGI}^{p,q}$ is within the interval of [0, 1]. In accordance with this, Ngwenyama et al. (1996) proposed three group consensus indicators for the group as: (i) group strong agreement quotient (GSAQ), a measure of the level of agreement in the group, (ii) group strong disagreement quotient (GSDQ), a measure of the level of disagreement in the group, and (iii) group strongest disagreement indicator (GSDI), a measure of the breadth of opinions in the group. In addition, they also suggested three individual consensus indicators for the individuals: (i) individual strong agreement quotient (ISAQ), (ii) individual strong disagreement quotient (ISDQ), and (iii) individual strongest disagreement indicator (ISDI). These three measures estimate the position of each individual relative to the group. Two thresholds of agreement and disagreement, α and β, respectively, could be defined to judge the level of consensus of the group. If $\mathrm{AGI}^{p,q} \geq \alpha$ to be the status of strong agreement $\Gamma(p, q) = 1$; otherwise, $\Gamma(p, q) = 0$. $\mathrm{AGI}^{p,q} \leq \beta$ to be the status of strong disagreement $\Phi(p, q) = 1$; otherwise, $\Phi(p, q) = 0$. Based on the group preference, U_G, aggregated from individual preferences, u^k, $k = 1,\ldots, K$, Bryson (1996) simplified Ngwenyama et al.'s six formulae (1996) as follows.

$$\mathrm{GSAQ}_\alpha = \sum\nolimits_{p\in S} \Gamma(u^p, U_G)/K \text{ with the threshold value } \alpha \tag{4.12}$$

$$\mathrm{GSDQ}_\beta = \sum\nolimits_{p\in S} \Phi(u^p, U_G)/K \text{ with the threshold value } \beta \tag{4.13}$$

$$\mathrm{GSDI}_\beta = \min\left\{\mathrm{AGI}^{p,q} - \beta,\ \text{over all } (p, q)\right\} \text{with the threshold value } \beta \tag{4.14}$$

$$\mathrm{ISAQ}_\alpha^p = \sum\nolimits_{p\in S, p\neq q} \Gamma(u^p, u^q)/(K-1) \text{ with the threshold value } \alpha \tag{4.15}$$

$$\mathrm{ISDQ}_\beta^p = \sum\nolimits_{p\in S, p\neq q} \Phi(u^p, u^q)/(K-1) \text{ with the threshold value } \beta \tag{4.16}$$

$$\mathrm{ISDI}_\beta^p = \min\left\{\mathrm{AGI}^{p,q} - \beta,\ \text{over all } q \text{ in } S)\right\} \text{ with the threshold value } \beta \tag{4.17}$$

These indicators yield an estimation of the level of consensus existing in the group.

After obtaining the three group consensus indicators, Eqs. (4.12)–(4.14), the facilitator or chairman could assess the current level of consensus in the group and plan consensus-building strategies. One interesting issue is to identify the DM with the highest potential for consensus improvement. Besides, three individual consensus indicators, Eqs. (4.15)–(4.17), construct an individual consensus vector (ICV) for the pth individual preference as $\mathrm{ICV}^p = \left[\mathrm{ISAQ}_\alpha^p, -\mathrm{ISAQ}_\beta^p, \mathrm{ISDI}_\beta^p\right]$. Those whose ICVs are not dominated by any other individual preferences and who exceed a predetermined threshold value for thc strong agreement quotient. These individuals would have a high potential for consensus-building around their own positions (Ngwenyama et al. 1996). Individuals with a dominated vector may be labeled outliers and asked to adjust their preferences or be expelled from the group. Another option for identifying problematic objectives is by checking the values of existing GSAQ_α and GSDQ_β. The removal of problematic individuals could produce a higher level of group consensus, increasing GSAQ_α and/or decreasing GSDQ_β, which could be verified as the right action by both values.

The thresholds, α and β, are the critical issues for the judgment. Ngwenyama et al. (1996) and Bryson (1996) both suggested the angle θ, between two preference vectors, 10 and 15 degrees for the thresholds. Nevertheless, Ngwenyama et al. (1996) considered the cosine function, $\cos(\theta)$, and Bryson (1996) took advantage of the function of $1-\sin(\theta)$ and other considerations as well. In spite of both functions being from the same concept, Ngwenyama et al. (1996) proposed different threshold values, a set of 0.9848 and 0.9662 corresponding to an angle of degrees 10 and 15, respectively. Because the study involved AHP with 1 to 9 ratio scales, they thought that the angles of 10 and 15 degrees could correspond to 1 and 1.5 scales, individually. Since the values of thresholds α and β are important to determine the consensus, it would be better to execute a sensitivity analysis to ensure adequate values.

While employing their approach to group TOPSIS, we observe that there are some possible expansions on the works of Ngwenyama et al. (1996) and Bryson (1996). The first one is on the choice of other types of similarity functions, e.g., Chen's similarity measure (1996). The second one involves the selection of an aggregation function, e.g., arithmetical mean (AM) or ordered weighted averaging operators. The third is related to the coverage of preference vectors. In group decision, the third dimension, DMs, is appended to the decision matrix, with alternatives and criteria. Our concentration could be generalized as the consensus on criteria and alternative. The former checks the deviations of the performance on any criterion by the group of members/individuals; the latter examines the discrepancy of the total performance of an alternative, with multiple criteria, by the group's members. The former is known as consensus at the criterion level, whereas the latter is known as consensus at the alternative level. The data from both can provide profound insights for consensus and would be quite useful in the process of achieving consensus. Moreover, Ma et al. (2014) suggested a three-level similarity measuring method of participant opinions: (i) assessment level for the similarity about any criterion, (ii) criterion level for each criterion's similarity to the criteria set, and (iii) problem level for the similarity between two opinions by participants. It seems that the second level could not provide effective information for decision support.

Let's illustrate the consensus indicators for management through an example.

Example 4.2 A manager's recruitment and selection problem [continued I]

The recruitment and selection problem from Example 4.1 (Shih et al. 2007) is selected to demonstrate the consensus reaching process in group TOPSIS. We re-organize the decision data shown in Table 4.9, from Tables 4.4 and 4.6.

Using an arithmetical mean for group preference U_G while following Ngwenyama et al. (1996) and Bryson's (1996) use of geometric mean for AHP preferences, we calculate the six consensus indicators, Eqs. (4.12)–(4.17), shown in Table 4.10. Although there are nine criteria to be evaluated by group TOPSIS, only two subjective criteria, panel interview and 1-on-1 interview, are selected for consensus information, since there is no difference for the performance measures on objective criteria. In addition to the consensus at the criteria level, we also compute the consensus at the alternative level, processing the relative closeness of alternatives, which is the basis for ranking in TOPSIS.

From the standpoint of the group, the results have a high degree of agreement, i.e., all $\text{GSAQ}_\alpha = 1$ and $\text{GSDQ}_\beta = 0$ at the levels $\alpha = 0.9848$ and $\beta = 0.9662$. Using the values of GSDI_β, a measure of the breadth of opinions in the group, the consensus at alternative level is better than those on criterion level. From an individual aspect, the consensus at the criteria level also has a high level of agreement. However, at the alternative level, the ICV $\left[\text{ISAQ}_\alpha^p, -\text{ISAQ}_\beta^p, \text{ISDI}_\beta^p\right]$ reveals that the preferences of DM#3 are dominated by DM#1 and DM#2. This indicates that the consensus could be improved if we force DM#3 out of the group should a higher degree of consensus being necessary. Otherwise, the group could accept the results as the final decision. Here we make an extra computation after forcing DM#3 out of the group, with the corresponding outcomes presented in Table 4.11. The value of $\text{AGI}^{p,q}$, the similarity among DMs, indeed improves with a higher degree of agreement. At the alternative level, the ICV declares that the preferences of DM#4 are dominated by DM#1's and DM#2's. Though we could do so, it would have little meaning to delete one to reach a higher level of consensus. In addition, at the criteria level, we like to know which evaluation is away from the mean and whether this indicates opportunities to improve consensus. Let's see the performance deviation from its AM on criterion panel interview in Table 4.12. The last row and the rightmost column of the table indicate the maximum deviations among the DMs and alternatives, individually. It is evident that DM#3 scores Candidate E with the largest deviation from the mean value, with the value of 17.5, and thus has been asked to make a change. Suppose that he/she would like to adjust the score from 50 to 60, and the new consensus indicators presented in Table 4.13. Comparing the ICVs of Tables 4.10 and 4.13 on panel interview, the latter generates much closer values. We could infer that the group unanimously agreed on the modified judgment.

Table 4.9 Decision matrix on subjective criteria and relative closeness for 17 candidates by four DMs

Performance evaluation		Candidate																
		A	*B*	*C*	*D*	*E*	*F*	*G*	*H*	*I*	*J*	*K*	*L*	*M*	*N*	*O*	*P*	*Q*
Panel interview	DM#1	80	65	90	65	75	80	65	70	80	70	50	60	75	80	70	90	80
	DM#2	85	60	80	55	75	75	70	75	95	75	62	65	80	75	75	92	70
	DM#3	75	70	80	68	50	77	65	75	90	68	60	50	65	80	65	85	75
	DM#4	90	60	90	62	70	75	67	82	90	65	65	45	70	75	60	88	70
	AM	82.50	63.75	85.00	62.50	67.50	76.75	66.75	75.50	88.75	69.50	59.25	55.00	72.50	77.50	67.50	88.75	73.75
1-on-1 interview	DM#1	75	75	85	70	80	80	70	60	85	75	60	65	75	70	65	95	85
	DM#2	80	70	85	60	80	85	60	65	85	80	65	75	80	72	70	90	75
	DM#3	70	77	90	72	55	82	72	67	85	78	65	60	75	70	70	80	80
	DM#4	85	70	95	72	75	75	75	85	92	70	70	50	75	75	65	90	75
	AM	77.50	73.00	88.75	68.50	72.50	80.50	69.25	69.25	86.75	75.75	65.00	62.50	76.25	71.75	67.50	88.75	78.75
Relative closeness	DM#1	0.5700	0.4089	0.8178	0.4716	0.5439	0.6717	0.4517	0.4681	0.7951	0.4944	0.0927	0.2101	0.5665	0.6050	0.4263	0.9219	0.6956
	DM#2	0.6548	0.3328	0.7082	0.3011	0.5407	0.6422	0.3264	0.4491	0.8943	0.5266	0.1689	0.3130	0.5972	0.5258	0.4296	0.9268	0.4616
	DM#3	0.5108	0.5528	0.8170	0.5190	0.2168	0.7137	0.4587	0.5093	0.8960	0.5372	0.2395	0.1280	0.5082	0.5812	0.4129	0.7961	0.6397
	DM#4	0.7947	0.4078	0.9216	0.4652	0.5478	0.6075	0.5251	0.7809	0.9444	0.4454	0.4149	0.0545	0.5561	0.6076	0.3445	0.9094	0.5563
	AM	0.6326	0.4256	0.8161	0.4392	0.4623	0.6588	0.4405	0.5519	0.8825	0.5009	0.2290	0.1764	0.5570	0.5799	0.4033	0.8886	0.5883

Note AM means the arithmetical mean of the performances of all four DMs

Table 4.10 Consensus indicators by four DMs

Consensus evaluation	Member	Group							Individual			Note
		$AGI^{p,AM}$	$\Gamma(p, AM)$	$GSAQ_{\alpha}$	$\Phi(p, AM)$	$GSDQ_{\beta}$	$AGI^{p,AM} - \beta$	$GSDI_{\beta}$	$ISAQ_{\alpha}^{p}$	$-ISDQ_{\beta}^{p}$	$ISDI_{\beta}^{p}$	
Panel interview	DM#1	0.9980	1	1	0	0	0.0320	0.0312	1	0	0.0275	(1) $\alpha = 0.9848$ (2) $\beta = 0.9659$ (3) p representing any DM (4) AM is the abbreviation of arithmetical mean
	DM#2	0.9977	1		0		0.0318		1	0	0.0258	
	DM#3	0.9971	1		0		0.0312		1	0	0.0258	
	DM#4	0.9979	1		0		0.0320		1	0	0.0278	
1-on-1 interview	DM#1	0.9985	1	1	0	0	0.0326	0.0307	1	0	0.0268	
	DM#2	0.9975	1		0		0.0315		1	0	0.0247	
	DM#3	0.9975	1		0		0.0315		1	0	0.0271	
	DM#4	0.9966	1		0		0.0307		1	0	0.0268	
Relative closeness	DM#1	0.9947	1	1	0	0	0.0287	0.0233	0.3333	0	0.0086	
	DM#2	0.9901	1		0		0.0241		0.3333	0	0.0010	
	DM#3	0.9893	1		0		0.0234		0	0	0.0010	
	DM#4	0.9893	1		0		0.0233		0	0	0.0061	

Table 4.11 Consensus indicators by three DMs (excluding DM#3)

Consensus evaluation	Member	Group							Individual			Note
		$AGI^{p,AM}$	$\Gamma(p, AM)$	$GSAQ_{\alpha}$	$\Phi(p, AM)$	$GSDQ_{\beta}$	$AGI^{p,AM} - \beta$	$GSDI_{\beta}$	$ISAQ_{\alpha}^{p}$	$-ISDQ_{\beta}^{p}$	$ISDI_{\beta}^{p}$	
Panel interview	DM#1	0.9982	1	1	0	0	0.0323	0.0319	1	0	0.0278	The α, β, p, and AM are defined the same as Table 4.10
	DM#2	0.9984	1		0		0.0325		1	0	0.0284	
	DM#4	0.9978	1		0		0.0319		1	0	0.0278	
1-on-1 interview	DM#1	0.9987	1	1	0	0	0.0268	0.0247	1	0	0.0268	
	DM#2	0.9980	1		0		0.0247		1	0	0.0247	
	DM#4	0.9967	1		0		0.0247		1	0	0.0268	
Relative closeness	DM#1	0.9941	1	1	0	0	0.0282	0.0244	0.5000	0	0.0086	
	DM#2	0.9930	1		0		0.0271		0.5000	0	0.0061	
	DM#4	0.9903	1		0		0.0244		0	0	0.0061	

Table 4.12 Deviation of the performance on panel interview for 17 candidates among DMs

Member	Candidate																	Max deviation
	A	*B*	*C*	*D*	*E*	*F*	*G*	*H*	*I*	*J*	*K*	*L*	*M*	*N*	*O*	*P*	*Q*	
DM#1	2.5	1.25	5	2.5	7.5	3.25	1.75	5.5	8.75	0.5	9.25	5	2.5	2.5	2.5	1.25	6.25	9.25
DM#2	2.5	3.75	5	7.5	7.5	1.75	3.25	0.5	6.25	5.5	2.75	10	7.5	2.5	7.5	3.25	3.75	10
DM#3	7.5	6.25	5	5.5	17.5	0.25	1.75	0.5	1.25	1.5	0.75	5	7.5	2.5	2.5	3.75	1.25	17.5
DM#4	7.5	3.75	5	0.5	2.5	1.75	0.25	6.5	1.25	4.5	5.75	10	2.5	2.5	7.5	0.75	3.75	10
Max deviation	7.5	6.25	5	7.5	17.5	1.75	3.25	6.5	8.75	5.5	9.25	10	7.5	2.5	7.5	3.75	6.25	

Table 4.13 Consensus indicators on panel interview by four DMs with modification

Member	Group							Individual			Note
	$AGI^{p,AM}$	$\Gamma(p, AM)$	$GSAQ_\alpha$	$\Phi(p, AM)$	$GSDQ_\beta$	$AGI^{p,AM} - \beta$	$GSDI_\beta$	$ISAQ_\alpha^p$	$-ISDQ_\beta^p$	$ISDI_\beta^p$	
DM#1	0.9981	1	1	0	0	0.0322	0.0319	1	0	0.0278	The α, β, p, and AM are defined the same as Table 4.10
DM#2	0.9978	1		0		0.0319		1	0	0.0277	
DM#3	0.9982	1		0		0.0323		1	0	0.0277	
DM#4	0.9979	1		0		0.0320		1	0	0.0278	

4.3.4 Madu's Approach

A control chart is a statistical process control tool used to determine if a manufacturing or business process is in a state of control or not. It is more appropriate to say that the control chart is the graphical device for statistical process monitoring (Wikipedia). It attempts to examine the difference between performance and its goal, i.e., the mean value. For variable data, the control chart for the sample means and sample range provides a powerful technique for analyzing process data. The ranges commonly represent three deviations of the sample means (Gryna 2000). The concept of control chart could be extended to investigate the opinions of the group's members away from its group means. This approach could be categorized at the criterion level since there is no real meaning to assess the data of relative closeness, ranking index, in TOPSIS. It is also considered as a static fashion, in which the data is taken only once in the decision process.

Madu (1994) employed a control chart approach to improve the quality of group decision making at the criterion level. He first utilized the coefficient of variation (CV) to measure the degree of disagreement among the members' evaluations of alternatives on each criterion. The CV is computed as the standard deviation divided by the mean. A high CV indicates a high degree of disagreement. The index of CV focuses on detecting sources of disagreement among members, especially on those alternatives with higher variability under any criterion. On a comparison basis, we know that if the group's judgment possesses higher diversity, the quality of the decision could be improved, at least as indicated by consistency.

The second part of Madu's method (1994) used a statistical control chart to measure the confidence of the decision. The GDM confidence limits could identify conflicts. Members in the group whose opinions fall outside the group's tolerance level are further studied for possible reduction of this variation. A discussion on the variation will be made so that either outliers change their preferences or the group agrees to the different viewpoint and re-executes preference elicitation. It could be possible that the group decides to exclude the preferences of outliers if the group operates in a dictatorial environment. The central part of control chart is to draw the chart with the mean, central line (CL), and the upper/lower control limits (UCL/LCL), and monitor the mean and range of normally distributed variables simultaneously if the data is under control or out of control (Gryna 2000). Some remedial actions will be recommended if the data is out of control limits. Here we further utilize $\overline{X}$-chart and R-chart, for variable data, to measure the quality of group decision with consensus on criteria. The former tries to measure central tendency of the data or preferences, and the later aims to determine the dispersion, i.e., range R, of the data. The shortcut formulas on sample averages $\overline{X}$ are (Gryna 2000):

$$\text{UCL} = \overline{\overline{X}} + A_2\overline{R} \tag{4.18}$$

$$\text{LCL} = \overline{\overline{X}} - A_2\overline{R} \tag{4.19}$$

where $\overline{\overline{X}}$ and $\overline{R}$ are the mean of $\overline{X}$ and R, respectively, and coefficient A_2 could be obtained from the appendix of this chapter.

Another shortcut formulas on sample ranges are as follows (Gryna 2000).

$$\text{UCL} = D_4\overline{R} \tag{4.20}$$

$$\text{LCL} = D_3\overline{R} \tag{4.21}$$

where $\overline{R} = \frac{\sum_{k=1}^{K}(R_{\max} - R_{\min})}{K}$ for K members, and coefficient D_3 and D_4 could be acquired from the appendix of this chapter.

It is worth noted that even though Madu (1994) successfully used control charts for a group decision with 17 individuals, we need to be careful in this use of control charts.

Let's use the same example to demonstrate the concept.

Example 4.3 A manager's recruitment and selection problem (continued II).

We would like to exhibit the consensus measures and reaching via control chart by the recruitment and selection problem from Example 4.1 (Shih et al. 2007). All the data is taken from Example 4.1, with four DMs. While Madu's work (1994) used AHP for preferences, this example uses group TOPSIS.

At the criterion level, panel interview and 1-on-1 interview are used for calculation of CVs for subjective criteria. Data normalization is accomplished using 100 for maximum values. Data, standard deviations (SDs), and CVs are presented in Table 4.14. From the displayed values, we know that there is high variation in ratings for candidates E, K, and L based on criterion panel interview, and on candidates E, H, and L on criterion based on 1-on-1 interview, all above 10%. These identify targets of adjustment to improve consensus. Since the control chart needs a large number of samples, our example could not meet the requirement with preference information from only four DMs. However, we would like to demonstrate the use of the chart. For the $\overline{X}$-chart on two criteria, the $\overline{\overline{X}}$ is equivalent to 0.7251 and 0.7484, and $\overline{R}$ equals to 0.1218 and 0.1241, respectively. According to Eqs. (4.18) and (4.19), the UCL and LCL for panel interview are 0.8139 and 0.6364, and for 1-on-1 interview are 0.8389 and 0.6579, individually. Two control charts are drawn in Figs. 4.3 and 4.4. In Fig. 4.3, it seems that DM#4 supplies a rather diverse evaluation away from the group mean on the criterion of panel interview. Figure 4.4 also shows that DM#4 provides a varied judgment. As a consequence, DM#4 should make an adaptation to reach a higher level of consensus. The detail suggestions on the evaluation of a specific candidate could also be found in the figures.

These two approaches only provide a general direction for consensus improvement. It would be valuable if the process could yield a recommendation on increasing or decreasing the judging values or keeping the status quo. Singh and Benyoucef (2013) proposed a soft consensus based GDM approach for consensus forming

Table 4.14 Normalized decision matrix and its coefficient of variations on two subjective criteria for 17 candidates by four DMs

Performance evaluation		Candidate																
		A	*B*	*C*	*D*	*E*	*F*	*G*	*H*	*I*	*J*	*K*	*L*	*M*	*N*	*O*	*P*	*Q*
Panel interview	DM#1	0.80	0.65	0.90	0.65	0.75	0.80	0.65	0.70	0.80	0.70	0.50	0.60	0.75	0.80	0.70	0.90	0.80
	DM#2	0.85	0.60	0.80	0.55	0.75	0.75	0.70	0.75	0.95	0.75	0.62	0.65	0.80	0.75	0.75	0.92	0.70
	DM#3	0.75	0.70	0.80	0.68	0.50	0.77	0.65	0.75	0.90	0.68	0.60	0.50	0.65	0.80	0.65	0.85	0.75
	DM#4	0.90	0.60	0.90	0.62	0.70	0.75	0.67	0.82	0.90	0.65	0.65	0.45	0.70	0.75	0.60	0.88	0.70
	AM	0.8250	0.6375	0.8500	0.6250	0.6750	0.7675	0.6675	0.7550	0.8875	0.6950	0.5925	0.5500	0.7250	0.7750	0.6750	0.8875	0.7375
	SD	0.0645	0.0479	0.0577	0.0557	0.1190	0.0236	0.0236	0.0493	0.0629	0.0420	0.0650	0.0913	0.0645	0.0289	0.0645	0.0299	0.0479
	CV	7.82%	7.51%	6.79%	8.91%	17.63%	3.08%	3.54%	6.53%	7.09%	6.05%	10.97%	16.60%	8.90%	3.72%	9.56%	3.36%	6.49%
1-on-1 interview	DM#1	0.75	0.75	0.85	0.70	0.80	0.80	0.70	0.60	0.85	0.75	0.60	0.65	0.75	0.70	0.65	0.95	0.85
	DM#2	0.80	0.70	0.85	0.60	0.80	0.85	0.60	0.65	0.85	0.80	0.65	0.75	0.80	0.72	0.70	0.90	0.75
	DM#3	0.70	0.77	0.90	0.72	0.55	0.82	0.72	0.67	0.85	0.78	0.65	0.60	0.75	0.70	0.70	0.80	0.80
	DM#4	0.85	0.70	0.95	0.72	0.75	0.75	0.75	0.85	0.92	0.70	0.70	0.50	0.75	0.75	0.65	0.90	0.75
	AM	0.7750	0.7300	0.8875	0.6850	0.7250	0.8050	0.6925	0.6925	0.8675	0.7575	0.6500	0.6250	0.7625	0.7175	0.6750	0.8875	0.7875
	SD	0.0645	0.0356	0.0479	0.0574	0.1190	0.0420	0.0650	0.1090	0.0350	0.0435	0.0408	0.1041	0.0250	0.0236	0.0289	0.0629	0.0479
	CV	8.33%	4.88%	5.39%	8.39%	16.42%	5.22%	9.39%	15.75%	4.03%	5.74%	6.28%	16.65%	3.28%	3.29%	4.28%	7.09%	6.08%

Note AM: arithmetical mean; SD: standard deviation; CV: coefficient of variation

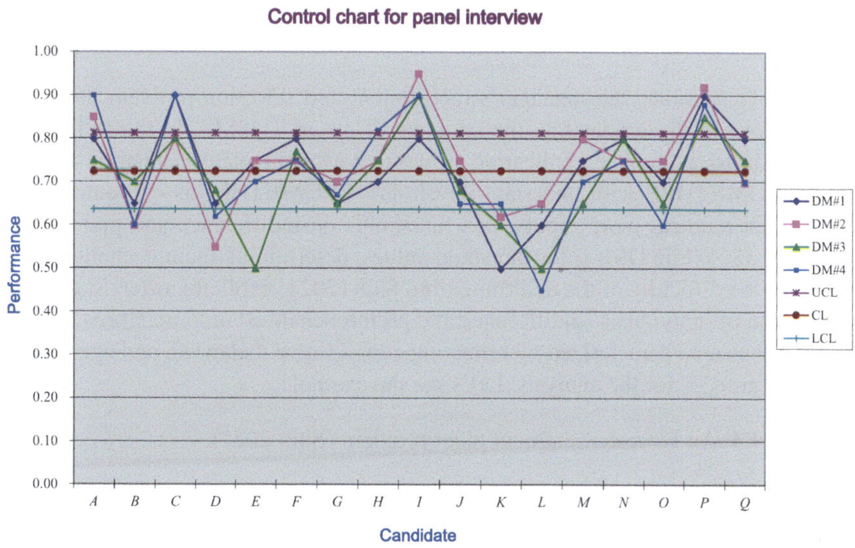

Fig. 4.3 Control chart for the criterion of panel interview

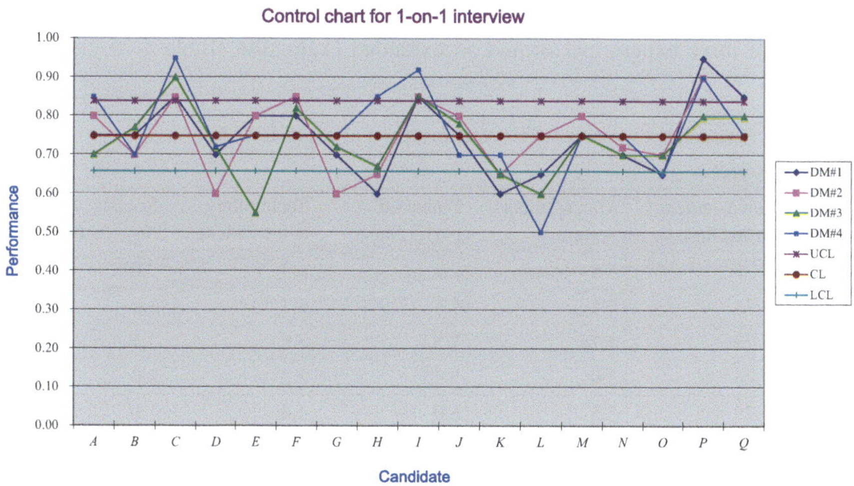

Fig. 4.4 Control chart for the criterion of 1-on-1 interview

among the supply chain partners regarding the preference values of various criteria for different alternatives. Regardless of the TOPSIS in crisp domain, all process is realized in the fuzzy domain. Interesting readers could refer to the contents of Dong et al. (2018) and Zhang et al. (2020) on consensus reaching outside the TOPSIS.

4.4 Consensus on Thresholds

In the above sections, the manager's recruitment and selection problem has been intensively used to demonstrate the contents of group TOPSIS. In the following, we will illustrate another application, a waste treatment plant evaluation by the group PROMETHEE with uncertain thresholds. PROMETHEE is one outranking method that requires more inputs, i.e., thresholds on indifference and preference, than TOPSIS. While DMs offer threshold values, determining them is challenging. To reduce the difficulty of the determination, Shih (2021) explicitly offers statistical aggregation of individuals' indifference and preference thresholds, and three typical combinations are characterized by reflecting conservative, balanced, and aggressive group preferences for the analysis. Let's see the example.

Example 4.4 An E-waste treatment plant problem (Shih 2021).

A project was initiated to evaluate waste electrical and electronic equipment (WEEE) treatment plants in Taiwanby the Recycling Fund Management Board (RFMB) of the Environmental Protection Administration. In order to rank and classify 15 local plants over five composite indices are: (i) environment protection, (ii) management system, (iii) financial performance, (iv) technology achievement, and (v) social responsibility, and the performance measures are shown in Table 4.15. A group of three experts performed an extended evaluation. Table 4.16 gives the

Table 4.15 Performance measures of E-waste treatment plants in Taiwan

Plant	Aspect				
	Environment protection	Management system	Financial performance	Technology achievement	Social responsibility
A	3.75	5.5	7.5	6.3	2.0
B	3.75	5.1	7.5	6.3	3.0
C	4.25	4.75	8.5	7.6	3.5
D	3.25	5.0	5.25	2.2	1.0
E	3.25	5.5	6.0	4.6	1.0
F	4.25	5.1	7.0	6.1	2.8
G	2.75	1.3	6.6	3.9	1.3
H	3.95	3.8	7.35	2.6	1.5
I	2.8	4.55	7.1	4.1	2.5
J	3.85	6.75	7.0	5.3	4.3
K	4.1	4.0	5.6	8.6	2.0
L	3.25	2.75	6.8	4.1	1.0
M	4.25	5.5	5.05	5.9	2.0
N	3.25	4.75	6.6	3.6	4.0
O	2.8	3.5	6.6	4.6	3.0

Table 4.16 Thresholds and criterion weights by the members of the group

Group member	Preference information		Environment protection	Management system	Financial performance	Technology achievement	Social responsibility
RFMB	Thresholds	Indifference q_j^1	0.15	0.20	0.25	0.30	0.20
		Preference p_j^1	0.35	0.50	0.60	0.80	0.40
	Criteria weight w_j^1		0.25	0.25	0.15	0.25	0.10
University	Thresholds	Indifference q_j^2	0.25	0.30	0.40	0.45	0.25
		Preference p_j^2	0.50	0.60	0.80	1.00	0.55
	Criteria weight w_j^2		0.4618	0.0898	0.0546	0.1853	0.2085
Industry	Thresholds	Indifference q_j^3	0.30	0.45	0.60	0.70	0.35
		Preference p_j^3	0.60	0.90	1.10	1.20	0.70
	Criteria weight w_j^3		0.2479	0.1528	0.4351	0.0613	0.1029

thresholds and weights on the five aspects by group members, i.e., q_j^*, p_j^*, and w_j^* on criterion j, $j = 1, \ldots, 5$, respectively. The study utilized PROMETHEE III, with interval coefficient $\gamma = 0.03$, to cover the imprecise measurement in the evaluation.

One core input to PROMETHEE is to determine the thresholds for the model. Although the determination is arduous, the group contribution could reduce the burden. We consider the thresholds are first roughly settled by individuals, and then group thresholds established based on these individual estimations. With simple statistical operations, i.e., maximum, minimum, and average, three typical group preferences were constructed: (i) conservative, (ii) aggressive, and (iii) balanced as the following Eqs. (4.22)–(4.24).

The minimum of indifference thresholds and the maximum of preference thresholds form a gradual slope, a conservative group preference or risk aversion, based on:

$$q_j^* = \mathrm{Min}(q_j^1, \ldots, q_j^K) \text{ and } p_j^* = \mathrm{Max}(p_j^1, \ldots p_j^K). \tag{4.22}$$

Here, $k = 1, \ldots, K$ for K members in the group.

The maximum of indifference thresholds and the minimum of preference thresholds form a sharp slope, in which the group preference is aggressive or risk seeking. The aggressive group preference is:

$$\begin{cases} q_j^* = \mathrm{Max}(q_j^1, \ldots, q_j^K) \text{ and } p_j^* = \mathrm{Min}(p_j^1, \ldots, p_j^K), \text{ if } \underline{p_j} \geq \overline{q_j}; \\ q_j^* = \mathrm{Max}(q^1, q^2, \ldots, q^K) = p_j^*, \text{ if } \underline{p_j} < \overline{q_j}. \end{cases} \tag{4.23}$$

where $\underline{p_j} = \mathrm{Min}(p_j^1, \ldots, p_j^K)$ and $\overline{q}_j = \mathrm{Max}(q_j^1, \ldots, q_j^K)$ on criterion j with K group members.

A balanced approach uses the central tendency of the thresholds, which takes the mean or average operator to describes the balanced group preference.

$$q_j^* = \mu_j^q \text{ and } p_j^* = \mu_j^p. \tag{4.24}$$

Here, $\mu_j^p = \sum_{k=1}^K \frac{p_j^k}{K}$ and $\mu_j^q = \sum_{k=1}^K \frac{q_j^k}{K}$, $k = 1, \ldots, K$ for K members in the group.

Table 4.17 lists the corresponding values by the three group preferences, from Eqs. (4.8)–(4.20), and group weights are the average of group members' values through agreement.

There are different versions of PROMETHEE (Brans and Vincke 1985). PROMETHEE II uses partial and complete rankings, while PROMETHEE III uses interval order for inputs, allowing rough expression of performance. TODIM is another outranking approach (Lee and Shih 2016). The impact of these different emphases led to different rankings, shown in Table 4.18 where PROMETHEE III rankings are given, following by PROMETHEE II rankings in parentheses.

These results demonstrate high degrees of similarity. Plant C was ranked first unanimously. There were clear differences across DMs, especially for the university

Table 4.17 Thresholds and criterion weights of the group with different group preferences

Group preference		Thresholds				
		Environment protection	Management system	Financial performance	Technology achievement	Social responsibility
Conservation Equation (4.18)	Indifference q_j^*	0.15	0.20	0.25	0.30	0.20
	Preference p_j^*	0.60	0.90	1.10	1.20	0.70
Aggression Equation (4.19)	Indifference q_j^*	0.30	0.45	0.60	0.70	0.35
	Preference p_j^*	0.35	0.50	0.60	0.80	0.40
Balance Equation (4.20)	Indifference q_j^*	0.2333	0.3167	0.4167	0.4833	0.2667
	Preference p_j^*	0.4833	0.6667	0.8333	1.0000	0.5500
Criteria weight of the group w_j^*		0.3199	0.1642	0.2132	0.1655	0.1371

Note w_j^* takes the average value of all members' weights of the group

Table 4.18 Ranks of plants by the single DM and the group by PROMETHEE with equal decision power

Plant	Single DM			Group				
	RFMB	University	Industry	Conservative	Aggressive	Balanced	Borda	TODIM
A	2 (4)	7 (7)	2 (4)	4 (5)	2 (5)	2 (5)	5	5
B	2 (5)	3 (6)	2 (3)	4 (4)	2 (4)	2 (4)	2	4
C	1 (1)	1 (1)	1 (1)	1 (1)	1 (1)	1 (1)	1	1
D	13 (14)	13 (14)	15 (15)	14 (14)	14 (14)	14 (14)	14	14
E	8 (8)	10 (10)	11 (13)	10 (10)	10 (11)	10 (10)	10	11
F	2 (3)	2 (2)	2 (5)	2 (2)	2 (2)	2 (2)	2	3
G	15 (15)	15 (15)	14 (14)	15 (15)	15 (15)	15 (15)	15	15
H	9 (10)	8 (8)	6 (6)	8 (8)	8 (8)	8 (8)	8	9
I	11 (11)	10 (12)	7 (8)	10 (11)	10 (10)	10 (11)	10	20
J	2 (2)	3 (3)	2 (2)	2 (3)	2 (3)	2 (3)	4	2
K	7 (7)	3 (5)	10 (10)	7 (7)	7 (7)	7 (7)	7	6
L	13 (13)	13 (13)	11 (12)	13 (13)	13 (13)	13 (13)	13	13
M	6 (6)	3 (4)	7 (9)	6 (6)	6 (6)	6 (6)	6	7
N	9 (9)	8 (9)	7 (7)	9 (9)	9 (9)	9 (9)	9	8
O	12 (12)	10 (11)	11 (11)	12 (12)	12 (12)	12 (12)	23	12

DM. Within the PROMETHEE III and PROMETHEE II ranks, there were few differences. The differences in emphasis between conservative, aggressive, and balanced emphases had practically no difference, and the BORDA and the TODIM rankings agreed with these three sets quite well. Additional sensitivity with respect to weight changes was conducted with little change in outcome.

4.5 Summary and Discussions

This chapter has exhibited the essence and processes of group decisions in TOPSIS. Groups could include multiple sources of knowledge and individual experience which can lead to better decisions. In the group process, the contents of consensus measures and reaching a semblance of agreement are also unveiled in detail by illustrative examples. TOPSIS is a simple technique and does not provide a means to elicit weights and robust judgment, particularly when dealing with subjective criteria and non-cardinal information, and a group of participants could contribute to these issues. Miscellaneous other supports from GDM can benefit when applying TOPSIS. CRP is an integrated process to reach a consensus, and it could help group TOPSIS decisions that are consistent and robust. The targets for processing consensus depend on the group preferences, and there are various targets to be chosen. Moreover, the variants, e.g., ordinal input and incremental analysis, in Chap. 3 can be extended to a group decision environment.

Appendix A

Constants for $\overline{X}$ and R charts

k	A_2	D_3	D_4	d_2
2	1.880	0	3.268	1.128
3	1.023	0	2.574	1.693
4	0.729	0	2.282	2.059
5	0.577	0	2.114	2.326
6	0.483	0	2.004	2.534
7	0.419	0.076	1.924	2.704
8	0.373	0.136	1.864	2.847
9	0.337	0.184	1.816	2.970
10	0.308	0.223	1.777	3.078

Note The table is from Gryna (2000)

References

Baucells, M., Sarin, R.K.: Group decisions with multiple criteria. Manage. Sci. **49**, 1105–1118 (2003). https://doi.org/10.1287/mnsc.49.8.1105.16400

Ben-Arieh, D., Easton, T.: Multi-criteria group consensus under linear cost opinion elasticity. Decis. Support Syst. **43**, 713–721 (2007). https://doi.org/10.1016/j.dss.2006.11.009

Bezerra, F., Melo, P., Costa, J.P.: Visual and interactive comparative analysis of individual opinions: a group decision support tool. Group Decis. Negot. **23**, 101–125 (2014). https://doi.org/10.1007/s10726-012-9330-6

Bryson, N.: Group decision-making and the analytic hierarchy process: exploring the consensus-relevant information content. Comput. Oper. Res. **23**, 27–35 (1996). https://doi.org/10.1016/0305-0548(96)00002-H

Chen, S.M.: New methods for subjective mental workload assessment and fuzzy risk analysis. Cybern. Syst. **27**, 449–472 (1996). https://doi.org/10.1080/019697296126417

Chen, S.M.: A new approach to handling fuzzy decision-making problems. IEEE Trans. Syst. Man Cybern. **18**, 1012–1016 (1988). https://doi.org/10.1109/21.23100

Chen, C.-T.: Extensions of the TOPSIS for group decision-making under fuzzy environment. Fuzzy Sets Syst. **114**, 1–9 (2000). https://doi.org/10.1016/S0165-0114(97)00377-1

Chu, T.-C.: Facility location selection using fuzzy TOPSIS under group decision. Int. J. Uncertian., Fuzz. and Knowl.-Based. Syst. **10**(6), 687–701 (2002). https://doi.org/10.1142/S0218488502001739

Colson, G.: The OR's prize winner and the software ARGOS: how a multijudge and multicriteria ranking GDSS helps a jury to attribute a scientific award. Comput. Oper. Res. **27**, 741–755 (2000). https://doi.org/10.1016/S0305-0548(99)00116-1

Cook, W.D.: Distance-based and ad hoc consensus models in ordinal preference ranking. Eur. J. Oper. Res. **172**, 369–385 (2006). https://doi.org/10.1016/j.ejor.2005.03.048

Dias, L., Clímaco, J.: ELECTRE TRI for groups with imprecise information on parameter values. Group Decis. Negot. **9**, 355–377 (2000). https://doi.org/10.1023/A:1008739614981

Dong, Y., Zha, Q., Zhang, H., Kou, G., Fujita, H., Chiclana, F., Herrera-Viedma, E.: Consensus reaching in social network group decision making: research paradigms and challenges. Knowl.-Based Syst. **162**, 3–13 (2018). https://doi.org/10.1016/j.knosys.2018.06.036

Forman, E., Peniwati, K.: Aggregating individual judgments and priorities with the analytic hierarchy process. Eur. J. Oper. Res. **108**, 165–169 (1998). https://doi.org/10.1016/S0377-2217(97)00244-0

Gryna, F.M.: Quality Planning and Analysis: From Product Development through Use, 4th edn. McGraw-Hill, New York (2000). ISBN: 9780070393684

Hitt, M.A., Miller, C.C., Colella, A.: Organizational Behavior, 4th edn. Wiley, Hoboken, New Jersey (2015). ISBN: 9781118809068

Huang, Y.S., Li, W.H.: A study on aggregation of TOPSIS ideal solutions for group decision-making. Group Decis. Negot. **21**, 461–473 (2012). https://doi.org/10.1007/s10726-010-9218-2

Hwang, C.L., Lin, M.J.: Group Decision Making under Multiple Criteria. Springer-Verlag, Berlin (1987). ISBN: 9783540171775

Labella, Á., Liu, Y., Rodríguez, R.M., Martínez, L.: Analyzing the performance of classical consensus models in large scale group decision making: a comparative study. Appl. Soft Comput. **67**, 677–690 (2018). https://doi.org/10.1016/j.asoc.2017.05.045

Lee, Y.S., Shih, H.S.: Incremental analysis for generalized TODIM. CEJOR **24**, 901–922 (2016). https://doi.org/10.1007/s10100-015-0427-2

Ma, J., Lu, J., Zhang, G.: A three-level-similarity measuring method of participant opinions in multiple-criteria group decision supports. Decis. Support Syst. **59**, 74–83 (2014). https://doi.org/10.1016/j.dss.2013.10.007

Madu, C.N.: A quality confidence procedure for GDSS application in multicriteria decision making. IIE Trans. **26**, 31–39 (1994). https://doi.org/10.1080/07408179408966605

Macharis, C., Brans, J.P., Mareschal, B.: The GDSS PROMETHEE procedure—A PROMETHEE-GAIA based procedure for group decision support. J. Decis. Syst. **7**, 283–307 (1998)

Ngwenyama, O.K., Bryson, N., Mobolurin, A.: Supporting facilitation in group support systems: techniques for analyzing consensus relevant data. Decis. Support Syst. **16**, 155–168 (1996). https://doi.org/10.1016/0167-9236(95)00004-6

Parkan, C., Wu, M.L.: Process selection with multiple objective and subjective attributes. Prod. Plann. Contr. **9**, 189–200 (1998). https://doi.org/10.1080/095372898234415

Pochampally, K.K., Gupta, S.M.: A business-mapping approach to multi-criteria group selection. IEEE Int. Symp. Electron. Environ. **2004**, 321–326 (2004). https://doi.org/10.1109/ISEE.2004.1299724

Robbins, S.P., Judge, T.A. (2019). Organizational Behavior, 18th Edn. Upper Saddle River, New Jersey: Pearson. ISBN: 9780134729329

Rodríguez, R.M., Labella, Á., Tré, G.D., Martínez, L.: A large scale consensus reaching process managing group hesitation. Knowl.-Based Syst. **159**, 86–97 (2018). https://doi.org/10.1016/j.knosys.2018.06.009

Shih, H.S.: An incremental analysis for MCDM with an application to group TOPSIS. Eur. J. Oper. Res. **186**, 720–734 (2008). https://doi.org/10.1016/j.ejor.2007.02.012

Shih, H.S.: A mixed-data evaluation in group TOPSIS with differentiated decision power. Group Decis. Negot. **25**, 537–565 (2016). https://doi.org/10.1007/s10726-015-9451-9

Shih, H.-S.: Treshold-enhanced PROMETHEE group decision support under uncertainties. Mathematical Problems in Engineering (2021). Article ID 5594074: 1-21. https://doi.org/10.1155/2021/5594074.

Shih, H.S., Shyur, H.J., Lee, E.S.: An extension of TOPSIS for group decision making. Math. Comput. Model. **45**, 801–813 (2007). https://doi.org/10.1016/j.mcm.2006.03.023

Shih, H.S., Lin, W.Y., Lee, E.S.: Group decision making for TOPSIS. In: Joint 9th IFSA World Congress and 20th NAFIPS International Conference, IFSA/NAFIPS 2001, July 25–28: Vancouver, Canada, pp. 2712–2717 (2001). https://doi.org/10.1109/NAFIPS.2001.943653

Shih, H.S., Wang, C.H., Lee, E.S.: A Multiattribute GDSS for Aiding Problem-solving. Math. Comput. Model. **39**, 1397–1412 (2004). https://doi.org/10.1016/j.mcm.2004.06.014

Singh, R.K., Benyoucef, L.: A consensus based group decision making methodology for strategic selection problems of supply chain coordination. Eng. Appl. Artif. Intell. **26**, 122–134 (2013). https://doi.org/10.1016/j.engappai.2012.03.013

Szmidt, E., Kacprzyk, J.: A consensus-reaching process under intuitionistic fuzzy preference relations. Int. J. Intell. Syst. **18**, 837–852 (2003). https://doi.org/10.1002/int.10119

Tsui, C.W., Tzeng, G.H., Wen, U.P.: A hybrid MCDM approach for improving the performance of green suppliers in the TFT-LCD industry. Int. J. Prod. Res. **53**, 6436–6454 (2015). https://doi.org/10.1080/00207543.2014.935829

Zhang, H., Zhao, S., Kou, G., Li, C.C., Dong, Y., Herrera, F.: An overview on feedback mechanisms with minimum adjustment or cost in consensus reaching in group decision making: Research paradigms and challenges. Information Fusion **60**, 65–79 (2020). https://doi.org/10.1016/j.inffus.2020.03.001

Chapter 5
TOPSIS for Decision Support

Decision support is used to help people make decisions, and decision support systems (DSS) involve accessing data and models to aid decisions (Wikipedia). This chapter will justify the use of decision support, introduce its configurations, and present some TOPSIS-supported DSSs to deal with real world problem-solving processes.

5.1 Necessity of Decision Support

In the process of problem-solving, DSS generally involves a combination of different techniques at different stages of the problem-solving process. In Sect. 1.2, we described the phases: (i) definition of the problem; (ii) construction of the model; (iii) solution of the model; (iv) validation of the model; and (v) implementation of the solution. While multi-criteria decision making (MCDM) techniques are powerful tools in dealing with problems, they cannot manage all issues in these phases. Hwang and Lin (1987) organized the related techniques of group decision making (GDM) to help make the necessary decisions. They arrange four phases for group participation: (i) extraction, (ii) exploration, (iii) selection, and (iv) execution. It is clear that MCDM can provide capability in the selection phase but has less capability in the other phases. This is why DSS is demanded for problem-solving.

TOPSIS, one MCDM technique, cannot handle all problem-solving process issues. In Chap. 2, we discussed two basic inputs: weights on criteria and the decision matrix, both needed to execute the TOPSIS algorithm. Beyond criteria weights and performance measures, there are numerous TOPSIS algorithm variants which were discussed in Chap. 3. These variants include normalization processes, methods to obtain criteria weights, distance functions, the selection of PIS and NIS, the relative closeness formula, order input, and weights on separation measures. These options could be set as elements of DSS functions or as a step for decision makers (DMs) input. In addition, other information could also be important, such as different data

H.-S. Shih and D. Olson, *TOPSIS and its Extensions: A Distance-Based MCDM Approach*, Studies in Systems, Decision and Control 447,
https://doi.org/10.1007/978-3-031-09577-1_5

forms, fuzzy distance formulas, and defuzzification formulas. All these options can be made available in TOPSIS-supported DSS systems.

5.2 DSS Configurations and Components

5.2.1 DSSs and GDSSs

A DSS is an information system that aids in business or organizational decision making. DSSs assist people in making decisions at the management, operations, and planning levels of a company. Decision-making aids can be totally electronic, human-powered, or a hybrid of the two (Wikipedia). From a technical viewpoint, it is assembled from four basic components: data, models, user interface, and knowledge. Each component has several variants depending on the nature of the management-decision situation and the specific technologies used for support (Turban et al., 2011). The data could be subjective, personal, or objective; it might be presented in other forms, e.g., figures or photos on a social network. The models, the core of the system, could be mathematical, statistical, analytical, or multi-criteria. The user interface is the space where interactions between humans and machines occur to assist users, analyzers, or DMs to control the system. It could be as simple as a spreadsheet, or as sophisticated as a well-designed web interface. The knowledge could include some tacit knowledge and some parameters set by experts. The system provides a means to integrate computer capabilities with human intelligence and analytic models to improve decision quality and make faster, smarter, better decisions. Applications and research areas can be found in Eom et al. (1998), Eom and Lee (1990), Arnott and Pervan (2005), Eom and Kim (2006), and Razmak and Aouni (2015).

In an organization, it is common to make a decision with a group of members. A group decision support system (GDSS) or group support system is designed to support the individual work of the members within a time-place paradigm. Different features may be required for synchronous vs. asynchronous communication, as well as local vs. distant communication, which are supported by various information technologies. GDSSs may include a number of tools and techniques designed to facilitate several tasks in group discussions, such as problem definition, explorations of issues, consensus building, group writing, activity coordination, knowledge sharing and accumulation, data and decision analysis (Adkins et al., 2003). From a broad sense of the classification, Holsapple and Whinston (1996) introduced multi-participant DSSs, supporting multiple participants engaged in a decision-making task, and classified them into two major types: GDSS and organizational DSS; negotiation support system is the overlap of the previous two. Multi-participant DSSs are the DSSs that aid multiple participants engaged in a decision-making task. While GDSS is intended for supporting choices made by a group, organizational DSS is established to help a DM a more elaborate infrastructure than a group, i.e., involving specialized roles, restricted communication patterns, differing authority levels. Besides the above systems, there are groupware, computer-mediated communication, computer-support cooperative work, and coordination technology, with information and communications technology, to facilitate the needed functions (Holsapple and Whinston, 1996; Turban et al., 2011). To deal with the variety of nuanced terms, we use the term GDSS to cover all of the above systems.

Pervan (1998) provided a review of the theory development and conceptual work of GDSSs. Fjermestad and Hiltz (2000) organized the methodology and results of GDSS studies into a four-factor framework consisting of contextual factors, intervening factors, adaptation factors, and outcome factors. The factors could explain the details of the field studies of GDSSs. Santos et al. (2016) examined GDSS features and their contributions to technology strategy decision making. They also analyzed how advances in features of GDSSs support improvements in different stages of the process. Observe that we also employ DSS as a general term for GDSS in this book.

To demosntrate the configuration of the GDSS, we present an example of recruitment and selection in human resource management (HRM) as follows.

Example 5.1 The configuration of a recruitment and selection GDSS (Shih et al., 2005).

A company wanted to implement a DSS for its recruitment and selection activities within the HRM department to reduce the burden of the process and kept precise personal data for ease of retrieval and access by the related DMs (Shih et al., 2005).

At the beginning of the recruitment and selection process, the HRM department announced the news of the vacancy, and many applications sent their personal data for the application. After screening out unfavorable applications and storing the relevant decision data, the personnel database was used to manage a pool of applicants' status. The knowledge relevant to the requirements of a certain task, as well as other supporting material, would be kept in the inner database. All decision techniques to support the analysis were covered by the model base. The GDSS engine allowed bidirectional communication between the system and its users, or the DM group, which included general members and the chairman or moderator. The engine served as the system's control center and connects the many components. It aided in the planning of the agenda, the acquisition and storage of information and knowledge, the streamlining of procedures, and the direction of decision-making activities for the system's integrated recruitment and selection process. When the recruitment and selection process were completed, the GDSS engine would generate a notification to all applicants about the decision. The detailed configuration and its elements are shown in Fig. 5.1. Furthermore, the system is implemented in a Microsoft Windows environment on networked-based personal computers (Shih et al., 2005).

5.2.2 *TOPSIS's Components*

From an MCDM viewpoint, the integration of various MCDM techniques can help GDSSs or DSSs support a wide range of decision situations and allow division

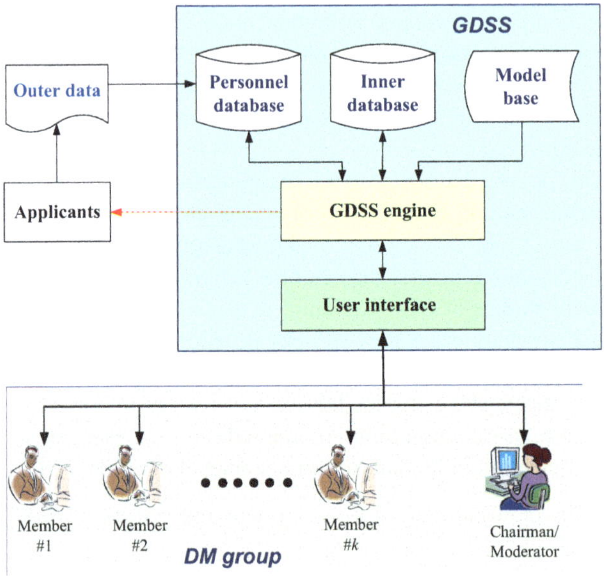

Fig. 5.1 Configuration of the recruitment and selection GDSS (Shih et al., 2005)

of decision-making tasks at particular stages (Bui, 1988). For TOPSIS-supported DSSs, we will illustrate their components. In Chap. 3, we presented numerous variants of TOPSIS for choice, including normalization processes, weights on criteria, distance functions, the selection of PIS and NIS, the relative closeness formula, order input, and weights on separation measures. Ploskas and Papthanasiou (2019) gave a comprehensive collection of the options for building a TOPSIS-supported DSS. They considered ten normalization techniques, three methods to calculate ideal and anti-ideal (nadir) solutions, fifteen distance metrics, five fuzzy distance metrics, and eight defuzzification techniques. They suggested TOPSIS and VIKOR (Opricovic and Tzeng, 2004) as MCDM models for DSS.

Table 3.1 gave formulas for eight normalization forms. Ploskas and Papthanasiou (2019) organized ten normalization techniques: (i) vector normalization (No. 4 in Table 3.1), (ii) linear sum normalization (No. 3 in the Table), (iii) linear max normalization (No. 1), (iv) linear max–min normalization (No. 2), (v) logarithmic normalization (No. 8), (vi) Marković method (2010), (vii) Tzeng and Huang method (2011), (viii) nonlinear normalization (No. 7 in Table 3.1), (ix) Lai and Hwang method (1994), and (x) Zavadskas and Turskis method (2008), in which the numbers in the parentheses are the orders listed in Table 3.1. In the above techniques, the Marković method, the Tzeng and Huang method, and the Lai and Hwang method are all linear normalization, which are the combinations of Nos. 1 and 2, respectively; however, the Marković method uses 1 to subtract the obtained value. The Zavadskas and Turskis method employs 1 minus the absolute values of No. 4.

Ploskas and Papthanasiou (2019) organized three methods to calculate distances to ideal and nadir solutions (PIS/NIS) are: (i) max–min values, as in traditional TOPSIS; (ii) absolute values, with all elements of criteria set to 1 for PIS and 0 for NIS; and (iii) fixed values based on maximum and minimum measures on each criterion, i.e., keeping the original values without normalization.

There are fifteen distance metrics for the separation measures, S_i^+ and S_i^-, and between alternative A_i and PIS/NIS, respectively, $i = 1,\ldots,m$, are arranged. The first three are Minkowski L_p metrics, which comprise: (i) Manhattan distance (sum of first powers) for $p = 1$, (ii) Euclidean distance (square root of sum of squares) for $p = 2$, and (iii) Tchebycheff distance (max differences) for $p = \infty$, respectively. These three are the most familiar metrics in TOPSIS (Olson, 2004). Please refer to Table 3.8 for the details. The fourth through the fifteen numbered 1 to 12, regardless of their concentration on statistical data, are listed in Table 5.1 with the corresponding formulae. Note that we only give the separation measures for benefit criteria in the table; the measures for the cost criteria will be in the reverse direction, i.e., $v_j^+ = \min_i v_{ij}$ and $v_j^- = \max_i v_{ij}$. It is worth mentioning that Mahalanobis distance (1936) is introduced at the end of the table, which incorporates the correlations among the criteria with a unit scale, and thus it does not require normalization on the original data (Vega et al., 2014). In addition, some explanations on the distances are presented in the note column of Table 5.1, taken from Wikipedia.

Numerous methods for calculating fuzzy distances have been developed (Zwick et al., 1987). Ploskas and Papathanasiou (2019) collected five fuzzy distances as the components of their DSS. For two triangular fuzzy numbers, $\tilde{m} = (m_1, m_2, m_3)$ and $\tilde{n} = (n_1, n_2, n_3)$, the distance between them is denoted by $d(\tilde{m}, \tilde{n})$. We tabulate these five distances in Table 5.2. On the other hand, the aim of defuzzification is to reduce the fuzzy set to a single crisp value, and it is the last step of fuzzy process (Van Leekwijck and Kerre, 1999). There are many methods for defuzzification, and Ploskas and Papathanasiou (2019) chose eight defuzzification methods, which are grouped in Table 5.3. A web-based DSS was established, with all of these options, to support either TOPSIS or VIKOR algorithms.

5.3 TOPSIS-Supported DSS Applications

There have been innumerable papers publishing DSSs employing TOPSIS. We review 11 applications to demonstrate some kinds of decisions often supported and the features of the systems developed.

5.3.1 Manufacturing Location Selection GDSS

Chi et al. (1996) described a DSS system that includes a database management system to store and retrieve information, which can be fed into a model base. Since most

Table 5.1 Extra used distances in TOPSIS

No	Distance measure	Note
1	Square Euclidean distance (sum of squares without square root): $S_i^+ = \sum_{j=1}^n (v_{ij} - v_j^+)^2, i = 1,\ldots, m,$ $v_j^+ = \max_i v_{ij}$ for benefit criteria $S_i^- = \sum_{j=1}^n (v_{ij} - v_j^-)^2, i = 1,\ldots, m,$ $v_j^- = \min_i v_{ij}$ for benefit criteria	Ploskas and Papathanasiou (2019)
2	Sørensen or Bray–Curtis distance: $S_i^+ = \frac{\sum_{j=1}^n \left\|v_{ij}-v_j^+\right\|}{\sum_{j=1}^n (v_{ij}+v_j^+)}, i = 1,\ldots, m,$ $v_j^+ = \max_i v_{ij}$ for benefit criteria $S_i^- = \frac{\sum_{j=1}^n \left\|v_{ij}-v_j^-\right\|}{\sum_{j=1}^n (v_{ij}+v_j^-)}, i = 1,\ldots, m,$ $v_j^- = \min_i v_{ij}$ for benefit criteria	(1) Sørensen (1948) or Bray and Curtis (1957) (2) To gauge the similarity of two samples
3	Canberra distance: $S_i^+ = \sum_{j=1}^n \frac{\left\|v_{ij}-v_j^+\right\|}{v_{ij}+v_j^+}, i = 1,\ldots, m,$ $v_j^+ = \max_i v_{ij}$ for benefit criteria $S_i^- = \sum_{j=1}^n \frac{\left\|v_{ij}-v_j^-\right\|}{v_{ij}+v_j^-}, i = 1,\ldots, m,$ $v_j^- = \min_i v_{ij}$ for benefit criteria	(1) From Lance and Williams (1966) (2) A weighted version of L_1 Manhattan distance
4	Lorentzian distance: $S_i^+ = \sum_{j=1}^n \ln(1 + \left\|v_{ij} - v_j^+\right\|), i = 1,\ldots, m,$ $v_j^+ = \max_i v_{ij}$ for benefit criteria $S_i^- = \sum_{j=1}^n \ln(1 + \left\|v_{ij} - v_j^-\right\|), i = 1,\ldots, m,$ $v_j^- = \min_i v_{ij}$ for benefit criteria	From Deza and Deza (2014)
5	Jaccard distance: $S_i^+ = \frac{\sum_{j=1}^n (v_{ij}-v_j^+)^2}{\sum_{j=1}^n (v_{ij})^2+\sum_{j=1}^n (v_j^+)^2-v_{ij}v_j^+}, i = 1,\ldots,$ $m, v_j^+ = \max_i v_{ij}$ for benefit criteria $S_i^- = \frac{\sum_{j=1}^n (v_{ij}-v_j^-)^2}{\sum_{j=1}^n (v_{ij})^2+\sum_{j=1}^n (v_j^-)^2-v_{ij}v_j^-}, i = 1,\ldots,$ $m, v_j^- = \min_i v_{ij}$ for benefit criteria	(1) Jaccard (1901) (2) To gauge the similarity of two samples

(continued)

Table 5.1 (continued)

No	Distance measure	Note
6	Dice distance: $S_i^+ = \frac{\sum_{j=1}^n (v_{ij}-v_j^+)^2}{\sum_{j=1}^n (v_{ij})^2+\sum_{j=1}^n (v_j^+)^2}, i = 1,\ldots, m,$ $v_j^+ = \max_i v_{ij}$ for benefit criteria $S_i^- = \frac{\sum_{j=1}^n (v_{ij}-v_j^-)^2}{\sum_{j=1}^n (v_{ij})^2+\sum_{j=1}^n (v_j^-)^2}, i = 1,\ldots, m,$ $v_j^- = \min_i v_{ij}$ for benefit criteria	(1) Dice (1945) (2) To gauge the similarity of two samples
7	Bhattacharyya distance: $S_i^+ = -\ln\left(\sum_{j=1}^n \sqrt{v_{ij}v_j^+}\right)^2, i = 1,\ldots, m,$ $v_j^+ = \max_i v_{ij}$ for benefit criteria $S_i^- = -\ln\left(\sum_{j=1}^n \sqrt{v_{ij}v_j^-}\right)^2, i = 1,\ldots, m,$ $v_j^- = \min_i v_{ij}$ for benefit criteria	(1) Bhattacharyya (1943) (2) The measures the similarity of two probability distributions
8	Hellinger distance: $S_i^+ = 2\sqrt{1 - \sum_{j=1}^n v_{ij}v_j^+}, i = 1,\ldots, m,$ $v_j^+ = \max_i v_{ij}$ for benefit criteria $S_i^- = 2\sqrt{1 - \sum_{j=1}^n v_{ij}v_j^-}, i = 1,\ldots, m,$ $v_j^- = \min_i v_{ij}$ for benefit criteria	(1) Hellinger (1909) (2) To quantify the similarity between two probability distributions
9	Matusita distance: $S_i^+ = \sqrt{2 - 2\sum_{j=1}^n v_{ij}v_j^+}, i = 1,\ldots, m,$ $v_j^+ = \max_i v_{ij}$ for benefit criteria $S_i^- = \sqrt{2 - 2\sum_{j=1}^n v_{ij}v_j^-}, i = 1,\ldots, m,$ $v_j^- = \min_i v_{ij}$ for benefit criteria	Matusita (1955)
10	Squared-chord distance: $S_i^+ = \sum_{j=1}^n (\sqrt{v_{ij}} - \sqrt{v_j^+})^2, i = 1,\ldots, m,$ $v_j^+ = \max_i v_{ij}$ for benefit criteria $S_i^- = \sum_{j=1}^n (\sqrt{v_{ij}} - \sqrt{v_j^-})^2, i = 1,\ldots, m,$ $v_j^- = \min_i v_{ij}$ for benefit criteria	From Prentice (1980)

(continued)

Table 5.1 (continued)

No	Distance measure	Note
11	Pearson χ^2 distance: $S_i^+ = \sum_{j=1}^{n} \frac{(v_{ij}-v_j^+)^2}{v_j^+}, i = 1,\ldots, m,$ $v_j^+ = \max_i v_{ij}$ for benefit criteria $S_i^- = \sum_{j=1}^{n} \frac{(v_{ij}-v_j^-)^2}{v_j^+}, i = 1,\ldots, m,$ $v_j^- = \min_i v_{ij}$ for benefit criteria	(1) Pearson (1900) (2) A measure how likely it is that any observed difference between sets of data arose by chance
12	Squared χ^2 distance: $S_i^+ = \sum_{j=1}^{n} \frac{(v_{ij}-v_j^+)^2}{v_{ij}+v_j^+}, i = 1,\ldots, m,$ $v_j^+ = \max_i v_{ij}$ for benefit criteria $S_i^- = \sum_{j=1}^{n} \frac{(v_{ij}-v_j^-)^2}{v_{ij}+v_j^+}, i = 1,\ldots, m,$ $v_j^- = \min_i v_{ij}$ for benefit criteria	From Pearson (1900)
13	Mahalanobis distance: $S_i^+ = \sqrt{(\mathrm{v} - v_j^+)^T \mathrm{S}^{-1}(\mathrm{v} - v_j^+)}, i = 1,\ldots,$ $m, v_j^+ = \max_i v_{ij}$ for benefit criteria $S_i^- = \sqrt{(\mathrm{v} - v_j^-)^T \mathrm{S}^{-1}(\mathrm{v} - v_j^-)}, i = 1,\ldots,$ $m, v_j^- = \min_i v_{ij}$ for benefit criteria where S is the covariance matrix of the distribution	(1) Mahalanobis (1936) (2) A measure of the distance between a point P and a distribution D (3) If the covariance matrix is the identity matrix, the Mahalanobis distance reduces to the Euclidean distance

human decisions involve multiple, conflicting objectives, they suggested the use of an analytic hierarchy process (AHP) to assess the relative preference weights of DMs being supported and TOPSIS to evaluate and compare options. They proposed a prototype GDSS to aid manufacturing location decisions. The criteria would be the location factors, and the database would provide data on each alternative's performance. The decision-making procedure in the DSS involves nine steps, following the TOPSIS algorithm we covered in Chap. 2:

Step 1: Ask users to use AHP to compare importance of location factors.

Step 2: Create the location decision matrix by identifying measures for each location factor.

Step 3: Normalize location data using the TOPSIS approach, by dividing each location's measure by the square root of the sum of the squared measures for all locations;

Table 5.2 Commonly used fuzzy distances for TOPSIS

No	Distance measure	Note
1	Chen's vertex method: $d(\tilde{m},\tilde{n}) = \sqrt{\frac{1}{3}[(m_1-n_1)^2+(m_2-n_2)^2+(m_3-n_3)^2}$	Chen (2000)
2	Cha and Jung's fuzzy distance operator: $d(\tilde{m},\tilde{n}) = \left\lvert\frac{m_1-n_1+m_2-n_2+m_3-n_3}{3}\right\rvert + \left\lvert\frac{m_1+m_3-2m_2}{2(m_3-m_1)} - \frac{n_1+n_3-2n_2}{2(n_3-n_1)}\right\rvert \times \left\lvert\frac{m_3-m_1+n_3-n_1}{4}\right\rvert$	Cha and Jung (2003)
3	Lin's fuzzy distance operator: $d(\tilde{m},\tilde{n}) = \sqrt{\frac{1}{6}\left[\sum_{i=1}^{3}(n_i-m_i)^2+(n_2-m_2)^2+\sum_{i=1}^{2}(n_i-m_i)(n_{i+1}-m_{i+1})\right]}$	Lin (2003)
4	Chen's similarity measure: $d(\tilde{m},\tilde{n}) = \frac{\sum_{i=1}^{3}\lvert m_i-n_i\rvert}{3}$	Chen (1996)
5	Chen and Hsieh's similarity measure: $d(\tilde{m},\tilde{n}) = \left\lvert\frac{m_1+4m_2+m_3}{6} - \frac{n_1+4n_2+n_3}{6}\right\rvert$	Chen and Hsieh (1999)

Step 4: For each criterion, convert normalized values into linear scales, with the best values are assigned 1.0 and the worst values assigned 0.0.

Step 5: Establish a weighted normalized decision matrix.

Step 7: Calculate relative closeness of each alternative to the ideal solution.

Step 8: Sort alternatives based on their relative closeness.

Step 9: Construct individual preference relations to location factors.

For group decisions, individual preference relations can be aggregated into group preferences. This process can lead to negotiation among DMs, where computer communication systems can aid intra-group discussion.

Chi et al. (1996) validated their systems with a manufacturing facility location decision. They conducted an empirical study comparing the decision outcomes of groups using the TOPSIS and knowledge-based system enhancements with those who only used the TOPSIS support. The first stage of the experiment was the generation of a short-list of the top five sites. The second stage was to have groups make a final selection.

All groups consisted of four people. From each of the three experimental groups, participants used TOPSIS as an individual DSS to select the top ten states from their personal perspective. Then the mediator used the knowledge-based system (KBS) as a GDSS to select five states, measuring group consensus. In each of the three control groups, participants also used TOPSIS individually to identify their ten preferred sites. Then the mediator used TOPSIS to select the top five sites. Group members were then asked to communicate to decide on the group's site selection.

Table 5.3 Commonly used defuzzification methods for fuzzy TOPSIS

No	Defuzzification method	Note
1	First of maxima (FOM): $\text{FOM}(A) = \min(\text{core}(A))$	(1) Selection of the smallest element of the core of A as the defuzzification value (2) The set of elements having the largest degree of membership in A is called the core of A
2	Last of maxima (LOM): $\text{LOM}(A) = \max(\text{core}(A))$	Selection of the greatest element of the core of A as the defuzzification value
3	Middle of maxima (MOM): $\lvert\text{core}(A)_{<\text{MOM}(A)}\rvert = \lvert\text{core}(A)_{>\text{MOM}(A)}\rvert$, if the core of A contains an odd number of elements $\lvert\text{core}(A)_{<\text{MOM}(A)}\rvert = \lvert\text{core}(A)_{>\text{MOM}(A)}\rvert \pm 1$, if the core of A contains an even number of elements	
4	Center of gravity (COG): $\text{COG}(A) = \frac{\sum_{x_{\min}}^{x_{\max}} x\mu_A(x)}{\sum_{x_{\min}}^{x_{\max}} \mu_A(x)}$	(1) Calculation of the COG of the area under the membership function $\mu_A(x)$ (2) The common and useful defuzzification technique
5	Mean of maxima (MeOM): $\text{MeOM}(A) = \frac{\sum_{x\in\text{core}(A)} x}{\lvert\text{core}(A)\rvert}$	(1) Calculation of the mean of all the elements of the core of A (2) A variant of COG method
6	Basic defuzzification distribution (BADD): $\text{BADD}(A) = \frac{\sum_{x_{\min}}^{x_{\max}} x\mu_A^{\gamma}(x)}{\sum_{x_{\min}}^{x_{\max}} \mu_A^{\gamma}(x)}$, where α is a free parameter in $[0, \infty]$. The parameter is used to adjust the method to the following special cases: $\text{BADD}(A) = \text{MeOS}(A)$, if $\gamma = 0$; $\text{BADD}(A) = \text{COG}(A)$, if $\gamma = 1$; $\text{BADD}(A) = \text{MeOM}(A)$, if $\gamma \to \infty$, where MeOS(A) is the mean of support of the core A	An extension of COG
7	Index center of gravity (ICOG): $\text{ICOG}(A, \alpha) = \frac{\sum_{x\in A_\alpha} x\mu_A(x)}{\sum_{x\in A_\alpha} \mu_A(x)}$	(1) Calculation of the COG of the fuzzy set that is obtained after putting all membership values below a certain threshold α equal to zero (2) $\alpha = 0$ in the proposed system
8	Bisector of area (BOA): $\sum_{x_{\min}}^{\text{BOA}} x\mu_A(x) = \sum_{\text{BOA}}^{x_{\max}} x\mu_A(x)$	The bisector is the vertical line that dividing fuzzy region into two sub-regions of equal area

Experimental outcome measures were group consensus, decision time, and communication productivity. Decision perceptions were measured by questionnaires asking about satisfaction with the decision process, decision confidence, decision task difficulty, group conflict, interest in using the system in the future, and interest in working with the assigned group in the future. The task context was for an international car company to select a new plant site to make an inexpensive sub-compact car. On a 1–5 scale, a 5 was the ideal (best) measurement. Table 5.4 gives the results as follows.

Table 5.4 shows two measures with some indication of a significant difference. The strongest point was that those using the KBS found the decision less difficult and experienced less conflict. To recap the decision support system features, we organize them as follows in Table 5.5.

Table 5.4 Proposed GDSS system advantage statistics (Chi et al., 1996)

Dependent variable	Units	KBS groups	Non-KBS groups	Significance
Group consensus	Matches of five	3.25	3.75	
Decision time	Minutes	28.67	29.33	
Communication productivity	Words/time	24.59	21.04	
Satisfaction with process	1–5 scale	3.67	3.50	
Agreement with final decision	1–5 scale	3.83	3.92	
Decision confidence	1–5 scale	3.67	3.92	
Task difficulty	1–5 scale	3.67	2.58	<0.05
Group conflict	1–5 scale	4.00	3.33	<0.10
Interest in future use	1–5 scale	3.25	3.75	
Interest in future group work	1–5 scale	3.92	3.58	

Table 5.5 Features of the manufacturing location selection GDSS (Chi et al., 1996)

Decision	Manufacturing site selection
Decision maker(s)	Generic
Data	Location database
Models	AHP TOPSIS
System components	KBS to identify group consensus Discussion Individual decision makers linked (communication)

Table 5.6 Features of the product development DSS (Byun and Lee, 2005)

Decision	Select rapid prototyping system
Decision maker(s)	65 surveyed for conjoint
Data	Questionnaire for conjoint Physical characteristics database
Models	Conjoint analysis for part design AHP for weights TOPSIS for selection

5.3.2 Product Development DSS

The Internet era has seen an explosion in computer system applications for business. Rapid prototyping is one of the product development processes aimed at quickly providing useful systems. Byun and Lee (2005) developed a DSS to aid in selecting rapid prototyping systems best suited to the end use. They identified typical decision-making criteria and used conjoint analysis to reflect user preferences. TOPSIS was used to rank available alternatives.

Conjoint analysis was used to structure the consumer preferences of 42 users to design a test part. The resulting part-worths were used to design a part. Once the part was designed, six factors were identified as important criteria: (i) accuracy, (ii) surface roughness, (iii) tensile strength, (iv) elongation, (v) part cost, and (vi) time to build.

AHP was used to estimate relative criteria importance. The first four criteria were physically measurable. Fuzzy data was obtained to get an opinion on cost and time. Linguistic ratings for cost and time were converted into triangular fuzzy numbers. Then TOPSIS was applied to identify ideal and nadir solutions. Closeness coefficients were then calculated and used to rank six commercially available rapid prototyping systems. The DSS developed incorporated a number of multi-criteria models, including (i) conjoint analysis to aid product design, (ii) AHP to estimate relative importance, and (iii) TOPSIS to rank order alternatives. We organize the features of the product development DSS are shown in Table 5.6.

5.3.3 Recruitment and Selection GDSS

Shih et al. (2005) proposed a GDSS with multiple criteria to help recruitment and selection processes, the first few activities of HRM. Due to the fact that e-recruitment on the web is the current trend for recruitment and selection processes, we can distinguish many activities within the processes. They can (i) generate a pool of candidates for the job; (ii) have applicants fill out application forms; (iii) use various selection techniques to identify viable job candidates; (iv) send one or more viable job candidates to their supervisor; (v) have the candidate(s) go through selection interviews; and (vi) decide which candidate(s) should be offered the job. To meet the requirements of the activities of the process, Shih et al. (2005) designed a prototype system in two phases, each having four steps, with different time/space occurrences. Table 5.7 lists the steps of the procedure, the corresponding decision techniques, conflict existence, and time/space occurrence.

The prototype system has a flexible structure that can cover different types of recruitment and selection processes. Through an example provided by the HRM department of a local chemical company in southern Taiwan, seeking an on-line manager Five DMs from different departments of the company are involved in the process. The selected criteria incorporate knowledge tests (i.e., language tests, professional tests, and safety rule tests), skill tests (professional skills and computer skills), and interviews (panel interviews and one-to-one interviews). Also, the results of personality tests (Holland code, DISC measurement, and leader adaptability) are kept for reference in the beginning to ensure the suitability of the applicants. The features of the proposed recruitment and selection GDSS are shown in Table 5.8 as follows.

5.3.4 *Investment Evaluation DSS*

Tansel and Yurdakuil (2010) reported a credibility-scoring DSS for Turkish banks to quickly evaluate the financial status of manufacturing firms in search of loans. The system consisted of applying fuzzy TOPSIS to financial ratio data. An industry credibility score and a firm credibility score were fed into a fuzzy TOPSIS model to obtain an industry rating and a firm rating to rank 22 Turkish manufacturers. The industry rating module considered: (i) real gross domestic product growth, (ii) capacity utilization rate, (iii) labor productivity, (iv) real growth in goods exports, (v) imports to exports, (vi) ratio of exports of goods to gross domestic product, and (vii) ratio of imports of goods to gross domestic product.

Screens were provided for users to input fuzzy assessments for weights. Data was gathered from reports for each of the 22 manufacturers' performances on the seven criteria. The system quickly generated TOPSIS credibility scores, which were used to rank order the 22 firms. The DSS used to assess individual firms had inputs for: (i) current ratio; (ii) net earnings/net sales; (iii) bank debt/equity; (iv) liquidity ratio; (v) total liabilities/total assets; (vi) operating profit/net sales; and (vii) net earnings/equity. The bank user entered the industry, obtaining the industry credibility score from the Industry Rating module. The user could enter upper and lower threshold values, and the TOPSIS model identified the firm's credibility score, as well as the recommended choice. The features of the credibility scoring DSS are presented in Table 5.9.

Table 5.7 Procedure and contents of recruitment and selection in two phases (Shih et al., 2005)

No	Description	Technique used	Conflict existence	Time/space occurrence	Note
Step 1	Definition of recruitment and selection requirements: (1.1) Establishment of requirements of recruitment, (1.2) Clarification of personnel requirement for the job, (1.3) Creation of a set of criteria, and (1.4) Determination of minimum achievement for each criterion	General	Group	Same/different	Phase I (preprocess): Steps 1–4
Step 2	Identification of selection criteria for the requirements: (2.1) Silent generation of criteria in writing, (2.2) Round-robin recording of criteria, (2.3) Serial discussion of the list of criteria, and (2.4) Voting for necessary criteria	Nominal group technique	Group	Same/different	

(continued)

Table 5.7 (continued)

No	Description	Technique used	Conflict existence	Time/space occurrence	Note
Step 3	Elicitation of weights of criteria by individuals: (3.1) Establishment of a reciprocal matrix for criteria' comparison, 3.2) Calculation of criteria' weights by pair-wise comparison, and (3.3) Check for consistency of priority of the matrix for himself/herself	AHP	Individual	Same/different	
Step 4	Consensus facilitation for group judgment: (4.1) Derivation of GPAI for agreement on criteria, and (4.2) Examination of GPSI for agreement on weights of criteria	Consensus indices (Ngwenyama et al., 1996)	Group	Same/different	
Step 5	Establishment of a pool of applicants: (5.1) Recruitment from various resources, and (5.2) Establishment of a database for keeping applicants' status	Data entry	No	Different/ different	Phase II (selection process): Steps 5–8

(continued)

Table 5.7 (continued)

No	Description	Technique used	Conflict existence	Time/space occurrence	Note
Step 6	Screening and evaluation of applicants along a timetable: (6.1) Rejection of applications by background investigation, (6.2) Elimination of applicants with low test achievement, (6.3) Removal of applicants through structured interview(s), (6.4) Notification of unsuitable applicants, and (6.5) Possession of ranking results of candidates	Criteria' thresholds and the ranking	Group	Different/ different	
Step 7	Accumulated evaluation of candidates by individual (7.1) Accumulation of evaluation on each criterion for candidates; (7.2) Construction of the normalized decision matrix; (7.3) Construction of the weighted normalized decision matrix; (7.4) Determination of PIS and NIS; (7.5) Calculation of the separation measure; (7.6) Calculation of the relative closeness to the PIS; (7.7) Ranking of the candidates	TOPSIS	Individual	Same/same	

(continued)

Table 5.7 (continued)

No	Description	Technique used	Conflict existence	Time/space occurrence	Note
Step 8	Selection of suitable candidates (8.1) Aggregation of the evaluation results from individuals; (8.2) Ranking of the candidates (8.3) Notification of the candidates, successful and unsuccessful	Borda function	Group	Same/same	

Table 5.8 Features of the recruitment and selection GDSS (Shih et al., 2005)

Decision	Recruitment and selection processes
Decision maker(s)	Generic, four DMs for the case study
Data	Personnel database, inner database, model base
Models	AHP TOPSIS Nominal group technique Borda function Consensus reaching
System component	A network-based PC system with web interfaces User's interfaces connected to the server end via active server page (ASP) at each step SQL database Electronic meeting system

Table 5.9 Features of the credibility scoring DSS (Tansel and Yurdakuil, 2010)

Decision	Bank assessment of manufacturer credibility Assessment of industry credibility
Decision maker(s)	Banks
Data	Database on industries Database on firms
Models	Fuzzy TOPSIS

5.3.5 *Green Product Design DSS*

Fears of global warming have radically changed the attitudes of many people, making green products highly attractive. Tian et al. (2018) presented a DSS based on AHP and TOPSIS to aid designers in assessing new product proposals for green refrigerators in China. Initial data was gathered from surveys of professional, managerial, and academic experts. The approach was to use AHP to structure the decision problem and articulate preferences over criteria. AHP yields criterion weights. Then TOPSIS was applied to rank-order design alternatives. The main criteria and sub-criteria in the corresponding parentheses are: (i) environmental properties (CFCs in mg/m^3, CO_2 in mg/m^3, SO_2 in mg/m^3, phosphorus in mg/liter, suspended solids in mg/liter, using noise in decibels, and product noise in decibels), (ii) energy properties (energy types, performance ratio, and rate of energy utilization), (iii) resources properties (stock utilization, material yield rate, toxic material proportion, and harmful material proportion), and (iv) economic properties (consumer's cost/production cost, and environmental cost/production cost). Weights were generated using pairwise comparisons in AHP. TOPSIS was applied and used to rank alternatives. An extensive sensitivity analysis was conducted. The features of the green product design DSS are presented in Table 5.10.

Table 5.10 Features of the green product design DSS (Tian et al., 2018)

Decision	Design green products
Decision maker(s)	Manufacturers
Data	Product components
Models	AHP Gray correlation TOPSIS

5.3.6 *Closed-Loop Supply Chain Network Design DSS*

Customers today have access to a much broader range of products, foods, and virtually every commodity and product due to the complex interactions of global supply chains. This leads to greater use of mass production, which in turn leads to waste product disposal issues. India faces grave challenges in the depletion of natural resources and management of excessive electronic-related waste. Closed-loop supply chain systems facilitate waste reduction through end-of-life strategies such as reuse, repair, refurbishing, remanufacturing, and recycling. The environment benefits through reduction in natural resource use and waste. Society benefits through job creation and more affordable prices. The use of closed-loop supply chain management enhances satisfaction of triple bottom line (economic, environmental, and social/human) factors. Darbari et al. (2019) proposed a closed-loop supply chain DSS to: (i) increase stakeholder involvement by integrating flow allocation and transportation decisions considering all three triple bottom line dimensions; (ii) a social objective function was defined to examine job opportunities, employee training, community service, and donations to non-government organizations; (iii) model uncertainty through fuzzy goals and constraints to support decision making; (iv) integrate AHP and TOPSIS to calculate stakeholder influence through weights; and (v) optimize closed-loop supply chain through a multi-objective model.

A closed-loop model included component suppliers feeding assembly centers, which created products and forwarded them to distribution centers serving the primary market. The primary market returned waste products after their useful lives to collection centers, which were donated to NGOs or recycled to scrap centers, with a portion of used products sent to dismantling centers, which could be resold to secondary markets or feed component refurbishing centers to reduce the demand for natural resource use by component suppliers. The DSS model implementing this system included a methodology integrating three multi-criteria fuzzy tools: (i) a hybrid AHP-TOPSIS model to determine relative weights for the three triple bottom line factors based on DM's preference input; (ii) a closed-loop supply chain network design configuration obtained from a fuzzy mixed-integer linear programming model with objectives of maximizing profit, minimizing carbon emissions, and maximization of social benefits; and (iii) a fuzzy goal programming model to generate efficient solutions for DM's consideration.

Stakeholders were identified as experienced supply chain managers, the head of the employee union, two environmental experts, a third-party regional manager,

Table 5.11 Features of the closed-loop supply chain network design DSS (Darbari et al., 2019)

Decision	Sustainable supply chain network design
Decision maker(s)	Supply chain managers
Data	Scrap production Recycling possibilities Resale opportunities Costs Emission rates
Models	AHP TOPSIS Multi-objective optimization model

a local municipal corporate official, an NGO worker, and two academicians by the system. The considered criteria were: use of recycled material, transportation network effectiveness, green practices, health and safety, minimizing waste, reverse logistics network effectiveness, adherence to legal norms, employee training and education, use of clean and upgraded technology, revenue contribution, and new employment. The DSS could enhance systematic closed-loop supply chain management, enhancing the accomplishment of triple bottom line objectives. The features of the DSS are described in Table 5.11 as follows.

5.3.7 Customer-Driven Product Design DSS

Our rapidly changing society includes complex and highly variable demands for products. Axiomatic design seeks to satisfy perceived needs through mapping functional requirements with design parameters (Suh, 1990). This approach is widely used in software and hardware development, as well as in machine design. Kumar and Tandon (2019) proposed a DSS to prioritize customer requirements, map product attributes to quantitative requirements, generate valid product design options, resolve conflicts, and recommend a design most likely to achieve desired customer satisfaction. The design procedure components were:

Step 1: Determine customer needs through surveys and interviews.

Step 2: Quality function development (QFD) is used to convert ambiguous requirements into actionable statements.

Step 3: Using AHP, determine relative requirement weights.

Step 4: Classification of requirements into functional (desired) and non-functional (quality attributes).

Step 5: Using axiomatic design, generate design possibilities.

Step 6: Using TOPSIS, choose the best design.

Table 5.12 Features of the safety inspection DSS (Kumar and Tandon, 2019)

Decision	Select the best product design
Decision maker(s)	Product managers
Data	Customer demand and expectations Components available
Models	QFD AHP TOPSIS

TOPSIS functions with the use of AHP to generate weights for criteria used in TOPSIS. Ideal and anti-ideal solutions were identified and relative closeness was calculated to rank alternatives. Kumar and Tandon (2019) used a simple product to demonstrate their DSS. In Table 5.12, we present the features of the customer-driven product design DSS.

5.3.8 Manufacturing System Reconfiguration DSS

Mabkhot et al. (2020) developed a semantic web DSS for monitoring and control of reconfigurability of manufacturing systems (RMSs). In order to coordinate between stakeholders, access distributed information, and encode knowledge for quicker process improvement, semantic web technologies were adopted. A general ontology framework to capture and structure knowledge related to risks and disruptions in manufacturing systems was adopted. The designed DSS had the following functions: (i) detect disturbances—order execution and performance indicators are monitored to identify deviations, and use semantic web rules to diagnose root causes and implement computer aided control; (ii) identify requirements by detecting the type of disturbance and resulting impact on production requirements in order to select candidate configurations; (iii) recommend configurations by using the ontology to infer a set of candidate solutions; (iv) select configurations, supported by TOPSIS analysis; (v) validate selections by displaying configurations to DMs; and (vi) memorization to record the history of disturbances identified, actions adopted to rectify them, and analyze success. The TOPSIS criteria and measures used to select appropriate configurations are displayed in Table 5.13.

In their TOPSIS model, priority was given to discrimination criteria seeking to react to the current situation and cope with new requirements. Once that requirement was met, priority was given to configurations successfully dealing with the situation with a minimum number of resources and maximum resource utilization. The suggested DSS to assist DMs with manufacturing system reconfiguration enables the ability to detect production problems, identify consequences and production requirements, recommend candidate configurations, and use TOPSIS to recommend the solution best matching preferences. The features of the RMS DSS are shown in Table 5.14.

Table 5.13 RMSs DSS criteria and measures (Mabkhot et al., 2020)

Criteria	Sub-criteria	Indicators
Discrimination	Resource	RUN (1 if used; 0 if not)
	Configuration	TS (1 if throughput ≥ required, 0 if not)
Operational	Resource	Number of operators Number of modules Module utilization
	Configuration	Earliness Lateness Waiting time Work in process Nearest to required Throughput
Strategic	Module	Number of additional modules Number of removed buffers Number of relocated buffers
	Buffer	Number of additional buffers Number of removed buffers Number of relocated buffers
	Operators	Number of additional operators Number of removed operators Number of relocated operators

Table 5.14 Features of the RMS DSS (Mabkhot et al., 2020)

Decision	Select of products Reconfigure manufacturing systems
Decision maker(s)	Manufacturing system managers
Data	Product components Measurements on system performance
Models	TOPSIS Case-based reasoning
Systems	Legacy software Recommendation system Ontology

5.3.9 Food Supply Chains DSS

Food supply chains are especially subject to bullwhip effects from the many uncertainties in the growing and processing transportation network. This leads to missed production schedules, lower customer service, excessive inventory and resulting waste, and difficult capacity planning decisions. Supplier assessment is very important. Ortiz-Barrios et al. (2020) presented a DSS for food supplier assessment using a hybrid approach with AHP, DEMATEL (decision-making trial and evaluation laboratory), and TOPSIS to evaluate pork supplier performance.

The methodology included five phases:

Phase 1: Experts are obtained to provide input to the AHP model.

Phase 2: The model is built based on scientific literature review, customer needs, and expert opinions.

Phase 3: AHP is applied, using expert pairwise comparisons.

Phase 4: DEMATEL is applied to discriminate between dispatchers and receivers and identify the strength of influence.

Phase 5: TOPSIS is applied to rank suppliers.

This DSS was applied in Columbia, including two experienced managers, two industrial engineering academicians, an administrative assistant with a great deal of experience, and a shareholder as experts. The operating main criteria and sub-criteria in the corresponding parentheses are: (i) quality (ISO 9001 certification, hygiene standards, and organoleptic properties for health protection), (ii) commercial profile (market experience, E-business, and market reputation), (iii) innovation (IT infrastructure and R&D capability), (iv) production capacity, (v) service level (post-sale service, flexibility, lead-time, order quantity, and guarantee duration), (vi) localization, and (vii) financial profile (purchase discount, bulk discount, payment deadline, price, and financial position). The DSS included screens to expedite AHP and DEMATEL inputs as well as display results in terms of clusters of sub-criteria and relative weights. The features of the food supply chain DSS are offered in Table 5.15.

Table 5.15 Features of the food supply evaluation DSS (Ortiz-Barrios et al., 2020)

Decision	Monitor pork supplier performance
Decision maker(s)	Food supply chain managers
Data	Supply chain elements Supplier inspections
Models	AHP DEMATEL TOPSIS

Table 5.16 Features of the safety inspection DSS (Zytoon, 2020)

Decision	Safety inspection priority and scheduling
Decision maker(s)	Government agents
Data	Safety inspection results
Models	Expert Delphi to identify criteria AHP TOPSIS

5.3.10 Safety Inspection DSS

Inspection agencies are responsible for implementing regulatory policies. Workplace safety has advanced a great deal in the past decades but continues to be difficult to regulate. Zytoon (2020) presented a prioritization model for inspectors to select the order in which inspections are conducted. The system consisted of expert input using Delphi analysis to identify criteria. AHP was used to generate weights and obtain scores. The TOPSIS-supported system was used in two cycles, the first with respect to main criteria and the second to obtain final firm priority indices for prioritization and scheduling. The system also included a database of inspections and scores. The following main criteria and sub-criteria are involved: (i) compliance history (number of previous violations, seriousness of previous violations, and employer response to non-compliance); (ii) complaint history (number of complaints, seriousness of complaints, and employer response to complaints); (iii) injury records (injury rate, severity rate, and major accidents); and (iv) industry characteristics (number of workers, vulnerability of workers, and employer response to complaints).

Sub-criteria and their scores were used in the first cycle of TOPSIS to rank firms with respect to relevant main criteria. In this first cycle, TOPSIS closeness scores and AHP weights were then used to rank firms. The features of the safety inspection DSS are presented in Table 5.16.

5.3.11 Automobile Components Procurement DSS

Component procurement is crucial in the automotive industry. Chen et al. (2021) proposed a two-stage DSS consisting of data envelopment analysis (DEA) followed by a TOPSIS model to screen suppliers. A fuzzy mixed integer programming model was then run to optimize order allocations. Suppliers were assessed on the main criteria and sub-criteria in the corresponding parentheses: (i) quality (certification system, security and information system, and random inspection of products), (ii) service (ability to complete production plan, timely supply chain response, on-time delivery and service, and guarantee for after-sales), (iii) technology (ability to exploit new products, product development schedule, and technological advance), and (iv) cost (cost control ability, price superiority, and contract execution cost). The features of the automobile components procurement DSS are shown in Table 5.17.

Table 5.17 Features of the automobile components procurement DSS (Chen et al., 2021)

Decision	Screen of suppliers Optimization of order allocations
Decision maker(s)	Automotive parts supply chain managers
Data	Parts requirements Supplicr database Experts for weights
Models	DEA TOPSIS Fuzzy mixed integer programming

5.4 Summary and Discussions

These eleven DSS systems supported by TOPSIS are only a random sampling of many. Table 5.18 gives a recapitulation of these applications:

TOPSIS has become widely used as a ranking tool in DSSs. It provides a straightforward multiple criteria computation for ranking, and it is easily integrated with other types of techniques of MCDM or GDM. Aside from the components of TOPSIS and fuzzy TOPSIS described by Ploskas and Papathanasiou (2019), we have shown 11

Table 5.18 DSS Examples

Authors	Published	Function	Supporting models
Chi et al.	(1996)	Manufacturing location GDSS	AHP for importance
Byun and Lee	(2005)	Computer system development DSS	Conjoint, AHP for importance
Shih et al.	(2005)	Recruiting and selection GDSS	NGT, AHP, Borda function
Tansel and Yurdakuil	(2010)	Bank loan assessment DSS	fuzzy TOPSIS
Tian et al.	(2018)	Green product assessment DSS	AHP for weights
Darbari et al.	(2019)	Supply chain network design DSS	AHP for weights
Kumar and Tandon	(2019)	Customer-driven product design DSS	Quality function deployment, AHP
Mabkhot et al.	(2020)	Manufacturing system reconfiguration DSS	Semantic web to identify problems
Ortiz-Barrios et al.	(2020)	Food supplier assessment DSS	AHP, DEMATEL for weights, influence
Zytoon	(2020)	Safety inspection DSS	Delphi, AHP
Chen et al.	(2021)	Automobile parts procurement DSS	DEA

applications where it has been used within a DSS to provide an effective problem-solving process. Seeing that the basis of ranking is a continuous score, this has also been used to provide a credibility score. The ability to implement TOPSIS through a spreadsheet, Excel, or web-based connection techniques makes it attractive as a component for a DSS in the future.

References

Adkins, M., Burgoon, M., Nunamaker, J.F.: Using group support systems for strategic planning with the United States Air Force. Decis. Support Syst. **34**, 315–337 (2003). https://doi.org/10.1016/S0167-9236(02)00124-0

Arnott, D., Pervan, G.: A critical analysis of decision support systems research. J. Inf. Technol. **20**, 67–87 (2005). https://doi.org/10.1057/palgrave.jit.2000035

Bhattacharyya, A.: On a measure of divergence between two statistical populations defined by their probability distributions. Bull. Calcutta Math. Soc. **35**, 99–109 (1943)

Bray, J.R., Curtis, J.T.: An ordination of the upland forest communities of Southern Wisconsin. Ecol. Monogr. **27**, 325–349 (1957). https://doi.org/10.2307/1942268

Bui, X.T.: Co-Op: A Group Decision Support System for Cooperative Multiple Criteria Group Decision Making. Springer-Verlag, Berlin (1988). ISBN: 9780387187532

Byun, H.S., Lee, K.H.: A decision support system for the selection of a rapid prototyping process using the modified TOPSIS method. Int. J. Adv. Manuf. Technol. **26**, 1338–1347 (2005). https://doi.org/10.1007/s00170-004-2099-2

Cha, Y., Jung, M.: Satisfaction assessment of multi-objective schedules using neural fuzzy methodology. Int. J. Prod. Res. **41**, 1831–1849 (2003). https://doi.org/10.1080/1352816031000074937

Chen, S.M.: New methods for subjective mental workload assessment and fuzzy risk analysis. Cybern. Syst. **27**, 449–472 (1996). https://doi.org/10.1080/019697296126417

Chen, C.T.: Extensions of the TOPSIS for group decision-making under fuzzy environment. Fuzzy Sets Syst. **114**, 1–9 (2000). https://doi.org/10.1016/S0165-0114(97)00377-1

Chen, J., Xu, Z., Gou, X., Huang, D., Zhang, J.: Automobile components procurement using a DEA-TOPSIS-FMIP approach with all-unit quantity discount and fuzzy factors. Technol. Econ. Dev. Econ. **27**, 311–352 (2021). https://doi.org/10.3846/tede.2020.13176

Chen, S.H., Hsieh, C.H.: Ranking generalized fuzzy number with graded mean integration representation. In: Proceedings of the Eighth International Conference of Fuzzy Sets and Systems Association World Congress, Taipei, vol. 2, pp. 551–555(1999)

Chi, S.C., Riordan, C.A., Benjamin, C.O.: The development and empirical study of a knowledge-based site-selection group decision support system. IEEE Int. Conf. Syst., Man, Cybern. **3**, 2053–2058 (1996). https://doi.org/10.1109/ICSMC.1996.565451

Darbari, J.D., Kannan, D., Agarwal, V., Jha, P.C.: Fuzzy criteria programming approach for optimizing the TBL performance of closed loop supply chain network design problem. Ann. Oper. Res. **273**, 693–738 (2019). https://doi.org/10.1007/s10479-017-2701-2

Deza, M.M., Deza, E.: Dictionary of Distances, 3rd edn. Springer-Verlag, Berlin (2014). ISBN: 9783662443415

Dice, L.R.: Measures of the amount of ecologic association between species. Ecology **26**, 297–302 (1945). https://doi.org/10.2307/1932409

Eom, S., Kim, E.: A survey of decision support system applications (1995–2001). J. Oper. Res. Soc. **57**, 1264–1278 (2006). https://doi.org/10.1057/palgrave.jors.2602140

Eom, H.B., Lee, S.M.: Decision support systems applications research: a bibliography (1971–1988). Eur. J. Oper. Res. **46**, 333–342 (1990). https://doi.org/10.1016/0377-2217(90)90008-Y

Eom, S.B., Lee, S.M., Kim, E.B., Somarajan, C.: J. Oper. Res. Soc. **49**, 109–120 (1998). https://doi.org/10.1057/palgrave.jors.2600507

Fjermestad, J., Hiltz, S.R.: Group support systems: a descriptive evaluation of case and field studies. J. Manag. Inf. Syst. **17**, 113–157 (2000). https://doi.org/10.1080/07421222.2000.11045657

Hellinger, E.: Neue Begründung der Theorie quadratischer Formen von unendlichvielen Veränderlichen. Journal Für Die Reine Und Angewandte Mathematik (in German) **136**, 210–271 (1909). https://doi.org/10.1515/crll.1909.136.210

Holsapple, C.W., Whinston, A.B.: Decision Support Systems: A Knowledge-Based Approach. West Publishing, St. Paul, MN (1996). ISBN: 978-0314069726

Hwang, C.L., Lin, M.J.: Group Decision Making under Multiple Criteria. Springer-Verlag, Berlin (1987). ISBN: 9783540171775

Jaccard, P.: Distribution de la flore alpine dans le Bassin des Dranses et dans quelques régions voisines. Bulletin De La Société Vaudoise Des Sciences Naturelles **37**, 241–272 (1901)

Kumar, P., Tandon, P.: A paradigm for customer-driven product design approach using extended axiomatic design. J. Intell. Manuf. **30**, 589–603 (2019). https://doi.org/10.1007/s10845-016-1266-2

Lai, Y.J., Hwang, C.L.: Fuzzy Multiple Objective Decision Making: Methods and Applications. Springer, Berlin (1994). ISBN: 9783540575955

Lance, G.N., Williams, W.T.: Computer programs for hierarchical polythetic classification ("similarity analyses"). Comput. J. **9**, 60–64 (1966). https://doi.org/10.1093/comjnl/9.1.60

Lin, Y.: Advanced Fuzzy Systems Design and Applications. Springer-Verlag, Berlin (2003). ISBN: 9783790815375

Mabkhot, M.M., Amri, S.K., Darmoul, S., Al-Samhan, A.M., Elkosantini, S.: An ontology-based multi-criteria decision support system to reconfigure manufacturing systems. IISE Transactions **52**, 18–42 (2020). https://doi.org/10.1080/24725854.2019.1597317

Mahalanobis, P.C.: On the generalised distance in statistics. In: Proceedings of the National Institute of Sciences of India, vol. 2, pp. 49–55 (1936). https://doi.org/10.1007/s13171-019-00164-5 (Reprinted)

Marković, Z.: Modification of TOPSIS method for solving of multicriteria tasks. Yugosl. J. Oper. Res. **20**, 117–143 (2010). https://doi.org/10.2298/YJOR1001117M

Matusita, K.: Decision rules, based on the distance, for problems of fit, two samples, and estimation. Ann. Math. Stat. **26**, 631–640 (1955). https://doi.org/10.1214/aoms/1177728422

Ngwenyama, O.K., Bryson, N., Mobolurin, A.: Supporting facilitation in group support systems: techniques for analyzing consensus relevant data. Decis. Support Syst. **16**, 155–168 (1996). https://doi.org/10.1016/0167-9236(95)00004-6

Olson, D.L.: Comparison of weights in TOPSIS models. Math. Comput. Model. **40**, 721–727 (2004). https://doi.org/10.1016/j.mcm.2004.10.003

Opricovic, S., Tzeng, G.H.: Compromise solution by MCDM methods: a comparative analysis of VIKOR and TOPSIS. Eur. J. Oper. Res. **156**, 514–529 (2004). https://doi.org/10.1016/S0377-2217(03)00020-1

Ortiz-Barrios, M., Miranda-De la Hoz, C., López-Meza, P., Petrillo, A., De Felice, F.: A case of food supply chain management with AHP, DEMATEL, and TOPSIS. J. Multi-Criteria Decis. Anal. **27**, 104–128 (2020). https://doi.org/10.1002/mcda.1693

Pearson, K.: On the criterion that a given system of deviations from the probable in the case of a correlated system of variables is such that it can be reasonably supposed to have arisen from random sampling. Philos. Mag. Ser. **5**(50), 157–175 (1900). https://doi.org/10.1080/14786440009463897

Pervan, G.P.: A review of research in Group Support Systems: leaders, approaches and directions. Decis. Support Syst. **23**, 149–159 (1998). https://doi.org/10.1016/S0167-9236(98)00041-4

Ploskas, N., Papathanasiou, J.: A decision support system for multiple criteria alternative ranking using TOPSIS and VIKOR in fuzzy and nonfuzzy environments. Fuzzy Sets Syst. **377**, 1–30 (2019). https://doi.org/10.1016/j.fss.2019.01.012

Prentice, I.C.: Multidimensional scaling as a research tool in quaternary palynology: a review of theory and methods. Rev. Palaeobot. Palynol. **31**, 71–104 (1980). https://doi.org/10.1016/0034-6667(80)90023-8

Razmak, J., Aouni, B.: Decision support system and multi-criteria decision aid: A state of the art and perspectives. J. Multi-Criteria Decis. Anal. **22**, 101–117 (2015). https://doi.org/10.1002/mcda.1530

Santos, C., Araújo, M., Correia, N.: Group support systems features and their contribution to technology strategy decision-making: a review and analysis. In: Schlick, C., Trzcieliński, S. (eds.): Advances in Ergonomics of Manufacturing: Managing the Enterprise of the Future, Advances in Intelligent Systems and Computing, vol. 490. Springer, Cham. https://doi.org/10.1007/978-3-319-41697-7_51

Shih, H.S., Huang, L.C., Shyur, H.J.: Recruitment and selection processes through an effective GDSS. Comput. Math. Appl. **50**(10–12), 1543–1558 (2005). https://doi.org/10.1016/j.camwa.2005.08.026

Sørensen, T.J.: A method of establishing groups of equal amplitude in plant sociology based on similarity of species content and its application to analyses of the vegetation on Danish commons. Kongelige Danske Videnskabernes Selskab, Biologiske Skrifter **5**, 1–34 (1948)

Suh, N.P.: The Principles of Design. Oxford University Press, New York (1990). ISBN: 9780195043457

Tansel, İÇ.Y., Yurdakul, M.: Development of a quick credibility scoring decision support system using fuzzy TOPSIS. Expert Syst. Appl. **37**, 567–574 (2010). https://doi.org/10.1016/j.eswa.2009.05.038

Tian, G., Zhang, H., Zhou, M.C., Li, Z.: AHP, gray correlation, and TOPSIS combined approach to green performance evaluation of design alternatives. IEEE Trans. Syst., Man, Cybern.: Syst. **48**, 1093–1105 (2018). https://doi.org/10.1109/TSMC.2016.2640179

Turban, E., Sharda, R., Delen, D.: Decision Support and Business Intelligence Systems, 9th edn. Prentice Hall, Upper Saddle River, New Jersey (2011). ISBN: 978-0136107293

Tzeng, G.H., Huang, J.J.: Multiple Attribute Decision Making: Methods and Applications. CRC Press, Boca Raton, FL (2011). ISBN: 9781439861578

Van Leekwijck, W., Kerre, E.E.: Defuzzification: criteria and classification. Fuzzy Sets Syst. **108**, 159–178 (1999). https://doi.org/10.1016/S0165-0114(97)00337-0

Vega, A., Aguarónb, J., García-Alcaraza, J., Moreno-Jiménez, J.M.: Notes on dependent attributes in TOPSIS. Procedia Comput. Sci. **31**, 308–317 (2014). https://doi.org/10.1016/j.procs.2014.05.273

Zavadskas, E.K., Turskis, Z.: A new logarithmic normalization method in games theory. Informatica **19**, 303–314 (2008). https://doi.org/10.15388/Informatica.2008.215

Zwick, R., Caristein, E., Budescu, D.V.: Measures of similarity among fuzzy concepts: a comparative analysis. Int. J. Approx. Reason. **1**, 221–242 (1987). https://doi.org/10.1016/0888-613X(87)90015-6

Zytoon, M.A.: A decision support model for prioritization of regulated safety inspections using integrated Delphi, AHP and double-hierarchical TOPSIS approach. IEEE Access **8**, 83444–83464 (2020). https://doi.org/10.1109/ACCESS.2020.2991179

Chapter 6
Sorting Problems by TOPSIS

Sorting decisions divide objects into ordered groups. This is a common decision problem in multi-criteria decision making (MCDM). In this chapter, we review published work involving sorting and classification work in Technique for Order of Preference by Similarity to Ideal Solution (TOPSIS), and propose a more generalized approach based on the concept of minimized deviations of the randomly selected alternatives to determine the cut-off values of the classification. Variants of ranking indices, distance functions, and weights are exploited in the search for a better result. An integrated sorting procedure is also illustrated. Examples are supplied to illustrate the advantages of the proposed approach.

6.1 TOPSIS for Sorting

A sorting problem is defined as an operation that divides alternatives into groups of alternatives according to some norms (Vincke 1992). The problem could be further divided into classification and clustering in data analytics (Olson and Delen 2008). Roughly speaking, if the number of groups is known a priori, the problem belongs to classification, and if the number of groups is not known a priori, to clustering. There are many practical applications in medicine, pattern recognition, human resource management, production systems management and technical diagnosis, marketing, environmental and energy management, and financial management and economics (Doumpos and Zopounidis 2002a). Moreover, with big data the problem is getting more and more important in business. For instance, sorting analysis could give retailers unprecedented, detailed insights into customer behavior, thus allowing them to improve the customer experience in a variety of ways: product recommendations, personalized search, customer support, and dynamic pricing. Thus, sorting is important, and we will show how TOPSIS, a distance-based compromise solution, can support sorting.

H.-S. Shih and D. Olson, *TOPSIS and its Extensions: A Distance-Based MCDM Approach*, Studies in Systems, Decision and Control 447,
https://doi.org/10.1007/978-3-031-09577-1_6

There were not many works on sorting problems when the TOPSIS method was introduced four decades ago. Wu and Olson (2006) utilized TOPSIS for in loan credit rating, using a cut-off value equivalent to its relative closeness obtained from the training data. Weights on criteria were attained from data by ordinal least squares (OLS) regression. In comparison with the results of decision tree algorithms, TOPSIS gave better fit to test data from Canadian banking loan cases with three metrics. Simulation was employed to examine the effects of perturbation data on the results. Chen et al. (2007) exploited distance measures to the fictitious ideal cases, similar to the TOPSIS concept, to classify water usage of Canada's 1285 municipalities. The data of three groups of representative cities, 12 cities in total, were analyzed with quadratic programming. The objective function minimized total error defined as deviations from group cut-off values, and the model identified cut-off values of groups and criteria weights. All Canadian municipalities were subsequently classified into three types: robust systems, low-risk systems, or high-risk systems with regard to water management. Observed that the cut-off value used here is the bounded value between two classes, while the term "cut-off value" also appears in Sect. 3.9 for benefit–cost evaluation.

Zhu et al. (2013) proposed C-TOPSIS classification for credit rating of applicants. From the training data, a 2-norm Euclidean distance function derived criteria weights, yielding a cut-off value for test applications. C-TOPSIS results on accuracy, computational efficiency, and interpretability were compared with those of seven popular data mining algorithms through Australian and German Credit Data Sets. Only the proposed method ranked among the top 3 in all three aspects to demonstrate its advantages. Sabokbar et al. (2016) presented a novel sorting method, TOPSIS-SORT, for evaluating Tehran environmental quality. Twenty-two districts were sorted into five groups, whose profiles and cut-off values on criteria were supplied by experts. Ouenniche et al. (2018) initiated an integrated in-sample and out-of-sample evaluation framework for bankruptcy prediction. Their key step was to find the upper and lower bounds of the TOPSIS score-based cut-off values, and they identified optimal values through non-linear search algorithms. Empirical results on a UK dataset of bankrupt and non-bankrupt firms showed outstanding prediction performance. Various normalization forms and distance metrics were available for choice in the TOPSIS non-parametric classifier.

de Lima Silva and de Almeida Filho (2020) considered two versions of cut-off value settings for sorting, TOPSIS-Sort-B based on the boundary profile from Sabokbar et al. (2016), and TOPSIS-Sort-C for the characteristic profile allowing upper and lower bounds on boundary profiles. Cut-off values or profiles on each criterion of both versions were delivered by experts to start algorithms. After exploring the case of world economic freedom in 180 countries, the similarity percentage or correct ratio of both versions was over 89%. de Lima Silva et al. (2020) went on to propose PDTOPSIS-Sort, a preference disaggregation TOPSIS approach for sorting in which non-linear programming was used to obtain boundary profiles and weights on criteria. The objective function was to minimize the sum of the error variables of the reference alternatives in their corresponding groups, following Doumpos and Zopounidis (2002b). The approach was applied to a 50-corporate bond classification problem with ten criteria given by nine reference alternatives in three classes. They claimed that the results were consistent with expert judgments.

Some published research has applied sorting to VIKOR (VlseKriterijumska Optimizacija I Kompromisno Resenje, VIKOR [from Serbian]; All/Multiple Criteria Optimization and Compromise Solution) (Opricovic 1998), a distance-based compromise MCDM method. Baccour (2018) gave a combined TOPSIS and VIKOR method called ATOVIC classification, and applied it to some UCI medical datasets. They separated datasets into reference and test sets, and used the core concept of VIKOR for classification, i.e., the relations among three indices. Five datasets were employed for comparison. Results affirmed that ATOVIC results had an edge over other data mining techniques. Demir et al. (2018) offered VIKORSORT for classifying 20 green suppliers into three groups. Using pre-assigned limit profiles on criteria, they utilized weighted distance to the ideal solution for the evaluation in classification. Ouenniche et al. (2021) expanded on Ouenniche et al.'s in-sample and out-of-sample evaluation framework (2018) for bankruptcy prediction using VIKOR. They considered the same index as Demir et al. (2018) for classification and obtained cut-off values are from the samples.

In summary, we organize the characteristics of the mentioned TOPSIS-based sorting methods in Table 6.1 by authors' names and year, type of cut-off values obtained, number of classes, and application. This table shows that core works on handling multiple cut-off values could be acquired by expert judgments, reference alternatives/cases, or training data. The latter two have the value of allowing the users to manage multiple classes. In addition, some authors claimed the sorting results were successful, but the results were rather dependent on the cases and are difficult to understand. Inspired by the concept of minimized total deviation, we will refer to Chen et al.'s work (2007) to obtain a reliable way to determine cut-off values. Observe that the relevant developments and applications are rather limited, and thus provide a gap for our study.

6.2 Proposed Sorting Approach

6.2.1 Determination of Cut-Off Values

The determination of cut-off values for multiple classes is a challenging task. The values are commonly obtained from expert judgments, reference alternatives/cases, or training data. The first approach, rough expert judgments, is useful, but it is difficult to verify the results in different environments, time periods, or different expert teams. The second and third approaches employ a similar concept in which the cut-off values can be procured from the alternatives in the different classes. Chen et al.'s work (2007) utilized case-based distance models with applications in water resource management. They tried to classify Canada's 1,285 cities into three classes, and they chose four representative cities in each class.

Table 6.1 Characteristics of TOPSIS-based sorting methods

No	Authors' names (year)	Type of cut-off values obtained	Number of classes	Application	Note
1	Wu and Olson (2006)	Training data	Two	Credit evaluation—Canadian banking loan	(1) Weights on criteria from OLS regression (2) Linear normalization on data (3) Evaluation by three metrics and two types of decision tree
2	Chen et al. (2007)	Reference alternatives/cases	Three	Water usage of Canadian 1,285 municipalities	(1) Case-based reasoning (2) A program with minimum overall square error to obtain weights and cut-off values
3	Zhu et al. (2013)	Training data	Two	Credit evaluation—Australian and German credit data sets	(1) A goal function of 2-norm Euclidean distance on two groups for obtaining weights on criteria (2) Linear normalization on data (3) Compare accuracy, complexity, and interpretability with 7 methods of data mining (4) C-TOPSIS
4	Sabokbar et al. (2016)	Expert judgments	Five	Environmental quality evaluation in Teheran	(1) Given weights and profiles on criteria (2) Linear normalization on data

(continued)

Table 6.1 (continued)

No	Authors' names (year)	Type of cut-off values obtained	Number of classes	Application	Note
5	Ouenniche et al. (2018)	Training data	Two	Bankruptcy prediction for London Stock Exchange	(1) Various subjective or objective weights for choice (2) Evaluation by two metrics and Mahalanobis distance (3) Four normalization schemes for choice (4) Compare with multiple discriminant analysis
6	Baccour (2018)*	Training data	Five/Three/Two	Medical disease diagnosis and others (UCI data sets)	(1) Three indices in VIKOR for classification purpose (2) Vector normalization on data (3) Nine values for the coefficient/weight of strategy of the majority of criteria (4) ATOVIC
7	Demir et al. (2018)*	Expert judgments	Three	Supplier evaluation	(1) One index in VIKOR for classification purpose (2) Linear normalization on data (3) VIKORSORT

(continued)

Table 6.1 (continued)

No	Authors' names (year)	Type of cut-off values obtained	Number of classes	Application	Note
8	de Lima Silva and de Almeida Filho (2020)	Expert judgments	Five	World economic freedom	(1) Given weights (2) Linear normalization on data (3) Three sorting procedures are compared (4) Discussion on the properties of sorting algorithms
9	de Lima Silva et al. (2020)	Reference alternatives	Three	Corporate bond evaluation	(1) Linear normalization on data (2) A program with minimum errors to obtain weights and cut-off values (3) Result consistent check by experts (4) PDTOPSIS-Sort
10	Ouenniche et al. (2021)*	Training data	Two	Bankrupt prediction for London Stock Exchange	(1) Equal weights for criteria (2) Evaluation by two metrics and Mahalanobis distance (3) Five values for the coefficient/weight of strategy of the majority of criteria (4) Compare with k-nearest neighbors algorithm

Note "*" utilization of the concept of VIKOR

Chen et al. (2007) first transformed the multi-criteria alternatives into a consequence data space, aggregating criteria performance and their corresponding weights. The proposed an algorithm that calculated the distance of each alternative from the fictitious ideal alternative, mimicking the separation measure to the ideal solution (PIS) in TOPSIS, and ranked the alternatives in ascending order. Once the number of classes L is identified, there should be $L - 1$ cut-off values to be determined. Then the performance of the representative alternatives with positive or negative tolerances, i.e., errors, will be imbedded into the constraint set to form the boundaries of the classes. The objective function is to minimize total squared errors by counting the deviations of the alternatives away from the boundaries. In addition, there are two types of deviations from the boundaries. For two neighboring classes, one class of the alternatives is with deviations less than a fixed cut-off value, and the other class of the alternatives is with deviations greater than the fixed cut-off value. The cut-off values and the corresponding weights on criteria could be successfully acquired by solving the quadratic programming problem.

This method provides an objective approach to multi-class problems. However, after examining other cases, we find Chen et al.'s approach (2007) difficult to follow without a detailed description. The first is that it is difficult to find multiple cut-off values, and their algorithm frequently stops with just one cut-off value. This means that the program searches for the boundary of only two classes. The second flaw is related to the resulting criteria weights. Almost half of the criteria weights are zeros, and the unbalanced weights are counter-intuitive. Based on these observations, we employ the concept of minimizing the sum of the squares of the residuals or deviations (Johnson and Bhattacharyya 2019). The proposed program seeks a cut-off value between the two neighboring classes in which the corresponding alternatives are selected. Furthermore, weight set selection is another critical issue. To avert the bias of decision makers' (DMs') preferences, we acquire the weights by applying some objective weighting approaches (Zardari et al. 2015). Since we do not have much information on the characteristics of the target dataset, the trained and test portions of the dataset could be utilized to examine which weighting approach is suitable for the dataset. The recommended procedure can be called a case-based approach. Following the input of various weight sets into the program, the results on the misclassified alternatives could be used to validate the effectiveness of the weight sets.

Suppose that there is a cut-off value R_g to be determined for the gth and the $(g + 1)$th class. We will randomly select t_g numbers of alternatives for one class, and t_h numbers of alternatives from the other class. α_g^i, $i = 1, \ldots, m$, is an upper-bound error for ith alternative in the former class, where $-1 \leq \alpha_g^i \leq 0$; β_g^i is a lower-bound error for for ith alternative in the latter class, where $0 \leq \beta_g^i \leq 1$. The weighted normalized performance measure of alternative i is from their ideal solution can be represented as $\sum_j w_j \left| r_{ij} - r_j^+ \right|$. The suggested programming for two neighboring classes, g and $g + 1$, is:

$$\text{Minimize } \boldsymbol{d} = \sum\nolimits_{i=1}^{t_g} \left(\alpha_g^i\right)^2 + \sum\nolimits_{i=1}^{t_h} \left(\beta_{g+1}^i\right)^2 \tag{6.1}$$

Subject to

$$\left(\sum_i w_j \left| r_{ij} - r_j^+ \right|\right) + \alpha_g^i \leq R_g,\ g = 1, \ldots, L - 1 \tag{6.2}$$

$$\left(\sum_j w_j \left| r_{ij} - r_j^+ \right|\right) + \beta_{g+1}^i \geq R_g, \tag{6.3}$$

$$-1 \leq \alpha_g^s \leq 0, \tag{6.4}$$

$$0 \leq \beta_{g+1}^s \leq 1, \tag{6.5}$$

$$R_g \leq 1, \tag{6.6}$$

$$\sum_i w_j = 1, \tag{6.7}$$

$$w_j \geq 0. \tag{6.8}$$

where r_{ij}, $i = 1, \ldots, m$ and $j = 1, \ldots, n$, are the normalized performance values from the dataset; $r_j^+ = \max_i(r_{ij})$ for the benefit criteria and $r_j^+ = \min_i(r_{ij})$ for the cost criteria.

After assigning a weight set to the above program, the cut-off value R_g is obtained, which is the boundary of classes g and $g + 1$. If there are more than two classes, we will select the alternatives from the next two neighboring classes, $g + 1$ and $g + 2$, and run the program again. The process continues until there is no need for an extra class. In this fashion, the cut-off values $R_1, R_2, \ldots, R_{L-1}$ for L classes are acquired. Note that the selected numbers of alternatives in the two classes, i.e., t_g and t_h, may not be equal.

6.2.2 *Suggested Sorting Procedure*

In the original TOPSIS, the ranking index is relative closeness C_i^*, which is a combination of two separation measures S_i^+ and S_i^-, using an n-dimensional Euclidean distance. Since we do not know the characteristics of the targeted dataset in advance, some variants in the TOPSIS algorithm could be considered in verification looking for a better result. Note that S_i^+ and S_i^- are opposite concepts in assessment. Based on the concept of Wu and Olson's data mining method (2006), we propose a more general procedure as follows.

Step 1: Gather data.

(1.1) Provide a classification decision matrix or dataset.

Depending upon the characteristics of the data, organized or unorganized, if the data is unorganized, the data will be statistically tested for significance. The selected decision matrix, with m alternatives or instances and q criteria or columns, will be linearly normalized, which means the data is divided by their maximal value for benefit criteria and their minimal value is divided by the data for cost criteria, as follows.

$$\boldsymbol{R} = \begin{array}{c} \\ A_1 \\ A_2 \\ \vdots \\ A_i \\ \vdots \\ A_m \end{array} \begin{array}{c} \begin{array}{ccccccc} X_1 & X_2 & \cdots & X_j & \cdots & X_n \end{array} \\ \begin{bmatrix} r_{11} & r_{12} & \cdots & r_{1j} & \cdots & r_{1n} \\ r_{21} & r_{22} & \cdots & r_{2j} & \cdots & r_{2n} \\ \vdots & \vdots & \cdots & \vdots & \cdots & \vdots \\ r_{i1} & r_{i2} & \cdots & r_{ij} & \cdots & r_{in} \\ \vdots & \vdots & \cdots & \vdots & \cdots & \vdots \\ r_{m1} & r_{m2} & \cdots & r_{mj} & \cdots & r_{mn} \end{bmatrix}_{m\times n} \end{array} \tag{6.9}$$

where $i = 1,\ldots,m$ and $j = 1,\ldots,n$.

(1.2) Count the number of groups.

Step 2: Calculate the weights.

Based on the contents of the normalized matrix or dataset, Eq. (6.9), many candidate weight sets could be acquired. Some ordinarily used methods include the entropy method, the standard deviation method, and the CRiteria Importance Through Intercriteria Correlation (CRITIC) method. In addition, equal weights could be adjoined for comparison purposes when there is no preference on the criteria.

Step 3: Determine the cut-off values.

Execute the program, Eqs. (6.1)–(6.8), to determine the cut-off values for multiple groups. If there are L classes to be segregated, $L - 1$ cut-off values need to be discovered. Run the similar programs $L - 1$ times, with the representative alternatives in two neighboring classes. Based on the inputs from different weight sets, there are different groups of cut-off values sets obtained from program solution.

Step 4: Execute classification/sorting.

(4.1) Rank alternatives by TOPSIS.

According to the TOPSIS ranking index, including relative closeness C_i^* or separation measures S_i^+ and S_i^-, of alternatives i, $i = 1,\ldots,m$, allocate all alternatives in descending order, respectively. Manhattan distance is considered as an option for the analysis.

(4.2) Sort the alternatives into groups.

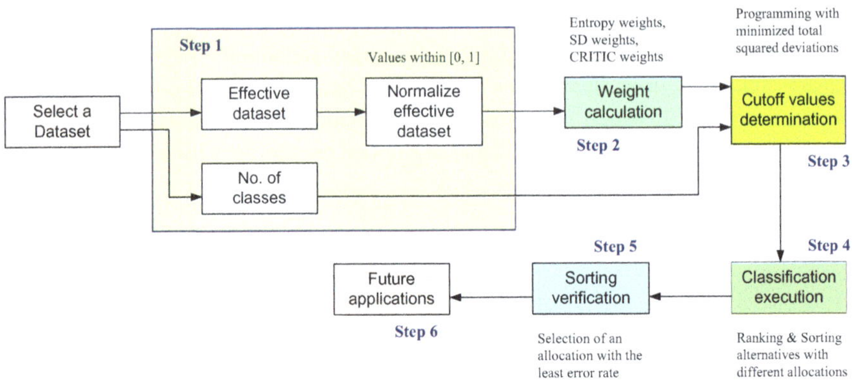

Fig. 6.1 Flowchart of the suggested sorting procedure

Consider the obtained cut-off values as the boundaries of the classes and compare the ranking indices of TOPSIS for the classification. Allocate all alternatives to the appropriate ordered classes. Here there are four groups of allocations from the four weight sets or allocations.

Step 5: Verify the classification.

Count the number of misplaced alternatives in each class, and enumerate the error rate of each allocation. Compare four groups of allocations and select one with the lowest error rate.

Step 6: Prospective application.

Since the choice is the classification with the least error rate, the corresponding cut-off values, with the corresponding weight set, could be used for classifying the datasets or decision matrixes with similar characteristics in the future.

The flowchart of the procedure is illustrated as follows (Fig. 6.1).

6.3 Analytic Results

6.3.1 Case Background

To demonstrate the effectiveness of the approach, we present a case classifying the level of economic freedom of countries in the world, which was originally provided by the Heritage Foundations (2021a, b at https://www.heritage.org/index/).

Economic freedom is the fundamental right of every human to govern his or her own labor and property. Individuals in an economically free society are free to work, create, consume, and invest as they see fit. Governments in economically free societies allow labor, capital, and goods to travel freely and refrain from restriction of liberty beyond what is required to safeguard and maintain liberty. Economic liberty leads to higher wealth. The Index of Economic Freedom shows how economic freedom correlates with a number of good social and economic outcomes. Economic freedom indicators are significantly linked to better communities, cleaner environments, more per capita wealth, human growth, democracy, and the abolition of poverty.

Economic freedom is measured using 12 quantitative and qualitative indicators that are divided into four major categories:

(i) rule of law (property rights, government integrity, judicial effectiveness);
(ii) government size (government spending, tax burden, fiscal health);
(iii) regulatory efficiency (business freedom, labor freedom, monetary freedom); and
(iv) open markets (trade freedom, investment freedom, financial freedom).

Each of these 12 economic freedom indicators is rated on a scale of 0 to 100. All indicators are considered equally significant in obtaining the positive effects of economic freedom. The overall score of a country is calculated by averaging these 12 indicators using equal weights. Countries that are considering economic reforms may discover huge benefits from the report (https://www.heritage.org/index/about).

We use the data of Year 2021 for the proposed analysis. There are 178 countries in the evaluation. The original process is equivalent to simple additive weighted (SAW) method with equal weights in MCDM. Based on the four cut-off values are 0.8, 0.7, 0.6 and 0.5, from expert judgments, each country is allocated into one of five classes.

6.3.2 Comparative Results

Weights of the 12 criteria from the four weight sets or allocations, entropy method, standard deviation method, CRITIC method, and equal weights, are listed in Table 6.2 (Step 2). One key step in the procedure is the determination of cut-off values in Step 3. Using the data from Year 2021, the four sets of cut-off values are illustrated in Table 6.3, which are from the averaged values of six randomized selected sets of alternatives. After obtaining cut-off values, we employ them for sorting countries (Step 4). In compliance with the ranking indices, relative closeness C_i^*, in the original TOPSIS, we also make use of separation measures to PIS/NIS S_i^+ and S_i^- for the evaluation despite the fact that the program concentrates on the distance to PIS. Their four error rates, counted by C_i^*, are presented in the fifth data column of Table 6.3, which are 27.53%, 22.47%, 46.63%, and 12.92%, respectively. Here equal weights provide the best result. The corresponding error rates for the two separation measures are also listed in the same table for comparison. It is clear that the ranking index obtained from the two separation measures would not yield good results. Besides the allocation by equal weights, entropy weights and standard deviation weights seem to provide the lower error rates (Step 5). Among the allocations, the CRITIC weight

Table 6.2 Weights on criteria form different weight sets (year 2021)

Weight set	Criteria											
	Property rights	Judicial effectiveness	Government integrity	Tax burden	Government spending	Fiscal health	Business freedom	Labor freedom	Monetary freedom	Trade freedom	Investment freedom	Financial freedom
Entropy weights	0.0881	0.1243	0.1286	0.0247	0.1043	0.1487	0.0472	0.0429	0.0309	0.0268	0.1192	0.1143
Standard deviation weights	0.0891	0.0954	0.1004	0.0560	0.0985	0.1195	0.0741	0.0692	0.0604	0.0562	0.0937	0.0875
CRITIC weights	0.0686	0.0791	0.0850	0.0796	0.1618	0.1425	0.0591	0.0698	0.0574	0.0462	0.0799	0.0710
Equal weights	0.0833	0.0833	0.0833	0.0833	0.0833	0.0833	0.0833	0.0833	0.0833	0.0833	0.0833	0.0833

Table 6.3 Cut-off values and error rates form different weight sets by Euclidean distance $L_{p=2}$ (year 2021)

Weight set	Cut-off values				Error rate			Note
	R_1	R_2	R_3	R_4	C_i^* (%)	S_i^+ (%)	S_i^- (%)	
Entropy weights	0.81790	0.71338	0.58320	0.48975	27.53	58.43	64.04	
Standard deviation weights	0.82276	0.72690	0.61195	0.52130	22.47	62.36	62.92	
CRITIC weights	0.80747	0.72981	0.63390	0.55585	46.63	46.63	80.34	
Equal weights	0.82439	0.73499	0.62891	0.53803	12.92	56.18	69.10	Reference

method yielded the results with the greatest error rates. Manhattan distance yielded error rates for different weight allocations are illustrated in Table 6.5 with equal weights and standard deviation weights fitting the data well.

To evaluate the effectiveness of the proposed procedure, we employ these weights and cut-off values to classify the datum of 2020 economic freedom (Step 6). The data is also from the website of the Heritage Foundations. The corresponding error rates are depicted in Table 6.4, which are slightly worse than the results of Year 2021's. The results show that entropy weights had the lower error rate, 29.44%, which are also higher than that of equal weights, 16.67%. Note that the error rate is close to the results of de Lima Silva and de Almeida Filho (2020), which are 89.44, 77.78, and 90.00% correct rates for directly using Year 2020 data, applying the three approaches.

Table 6.4 Error rates form different weight sets by Euclidean distance $L_{p=2}$ (year 2020)

Weight set	Error rate			Note
	C_i^* (%)	S_i^+ (%)	S_i^- (%)	
Entropy weights	29.44	68.89	64.04	Four cut-off values are shown in Table 6.3
Standard deviation weights	30.00	69.44	62.92	
CRITIC weights	45.56	55.56	61.67	
Equal weights	16.67	63.33	69.10	Reference

Table 6.5 Error rates form different weight sets by Manhattan distance $L_{p=1}$ (year 2021)

Weight set	Error rate			Note
	C_i^* (%)	S_i^+ (%)	S_i^- (%)	
Entropy weights	21.91	16.85	26.40	Four cut-off values are shown in Table 6.3
Standard deviation weights	11.24	9.55	25.28	
CRITIC weights	22.47	27.53	25.28	
Equal weights	9.55	10.11	21.35	Reference

Table 6.6 Error rates form different weight sets by Manhattan distance $L_{p=1}$ (year 2020)

Weight set	Error rate			Note
	C_i^* (%)	S_i^+ (%)	S_i^- (%)	
Entropy weights	20.56	15.00	32.22	Four cut-off values are shown in Table 6.3
Standard deviation weights	13.33	7.78	25.00	
CRITIC weights	22.22	23.89	31.11	
Equal weights	7.22	3.33	23.33	Reference

Another variant is employing Manhattan distance for the analysis. The corresponding results are presented in Table 6.5 for the year 2021 data. Standard deviation weights yield a better outcome with an error rate of 11.24%, which is still inferior to the 9.55% from equal weights. Using the distance measure to PIS also produces rather good results. By applying the same cut-off values to the data for the year 2020, the errors are displayed in Table 6.6. In general, standard deviation weights deliver better results with 13.33% error rate, which is inferior to 7.22% from equal weights. Here the error rates by counting the distance to PIS generate superior outcome, whose error rate reaches 3.33%, much better than de Lima Silva and de Almeida Filho's result (2020). It is apparent that Manhattandistance could provide lower error rate than Euclidean distance on this dataset. The phenomenon verifies that TOPSIS with Manhattan distance and SAW share the same characteristic (Hwang and Yoon 1981).

As discussed in the review, the difficulty for TOPSIS in classifying problems is in the determining cut-off values. The experts from the Heritage Foundations selected the four cut-off values are 0.8, 0.7, 0.6, and 0.5, respectively, for their SAW method. However, TOPSIS employs the value of relative closeness for ranking. The values are also considered as the cut-off values for classification. Hence, these four values for SAW are transformed into the corresponding cut-off values in Table 6.7, which vary depending on the given weight sets. The corresponding cut-off values and their error rates for the years 2021 and 2020 datasets are illustrated in the same table. Observe that these error rates are much better than our proposal, except for the CRITIC weights. Although it is presented for reference, equal weights produce the highest performance in terms of error rate, with 6.74% and 6.11%, respectively,

Table 6.7 Cut-off values and error rates form different weight sets by the original TOPSIS

Weight set	Cut-off values				Error rate		Note
	R_1	R_2	R_3	R_4	2021 (%)	2020 (%)	
Entropy weights	0.83526	0.69411	0.57285	0.46739	19.66	21.11	
Standard deviation weights	0.81887	0.70370	0.59587	0.49595	11.24	12.22	
CRITIC weights	0.80096	0.67058	0.60271	0.51714	60.11	55.00	
Equal weights	0.81826	0.71280	0.62066	0.52462	6.74	6.11	Reference

which is much better than the result of de Lima Silva and de Almeida Filho (2020). This evidence tells us that if we can precisely acquire the cut-off values, TOPSIS could be a good sorting tool for such type of datasets.

6.4 Summary and Discussions

This chapter proposes a procedure for TOPSIS to solve sorting problems. The major difference between TOPSIS, or any MCDM method, and data mining techniques is that the alternatives should be ranked first before the classification or clustering (Nemery 2008). Thus, an ordered group would be expected to be ranked first. This characteristic provides a unique feature for demonstrating ordered classes based on some ranking indices. Besides, searching for cut-off values among classes is difficult when using TOPSIS when applied to sorting problems, most of which are assigned by subjective values. We use the concept of the minimized square of deviations with alternative data to find adequate cut-off values. It is a case-based approach. For future applications on a similar dataset, error rates are employed to evaluate which weight set allocation under the variants of distance functions and ranking indices is suitable. In the economic freedom example, standard deviation weights with Manhattan distance could generally supply a better result. However, equal weights generate the best match to the original work. From these results, we know that the TOPSIS concept could be enhanced to provide a satisfactory result for sorting problems under the condition that the datasets are organized by the characteristics of the problems.

References

Baccour, L.: Amended fused TOPSIS-VIKOR for classification (ATOVIC) applied to some UCI data sets. Expert Syst. Appl. **99**, 115–125 (2018). https://doi.org/10.1016/j.eswa.2018.01.025

Chen, Y., Hipel, K.W., Kilgour, D.M.: Multiple-criteria sorting using case-based distance models with an application in water resources management. IEEE Trans. Syst., Man, Cybern. – Part A: Syst. Hum. **37**, 680–691 (2007). https://doi.org/10.1109/TSMCA.2007.902629

Demir, L., Akpınar, M.E., Araz, C., Ilgın, M.A.: A green supplier evaluation system based on a new multi-criteria sorting method: VIKORSORT. Expert Syst. Appl. **114**, 479–487 (2018). https://doi.org/10.1016/j.eswa.2018.07.071

Doumpos, M., Zopounidis, C.: Multicriteria classification and sorting methods: a literature review. Eur. J. Oper. Res. **138**, 229–246 (2002a). https://doi.org/10.1016/S0377-2217(01)00243-0

Doumpos, M., Zopounidis, C.: Multicriteria Decision Aid Classification Methods. Springer, New York (2002b). ISBN: 9781402008054

Heritage Foundation: 2021a index of economic freedom. https://www.heritage.org/index/ (2021a). Accessed November 3, 2021a

Heritage Foundation: About the index, 2021b index of economic freedom. https://www.heritage.org/index/about (2021b). Accessed November 3, 2021b

Hwang, C.L., Yoon, K.: Multiple Attribute Decision Making: A State-of-the-Art Survey. Springer-Verlag, Berlin (1981). ISBN: 9783540105589

Johnson, R.A., Bhattacharyya, G.K.: Statistics: Principles and Methods, 8th edn. Wiley, Hoboken, New Jersey (2019). ISBN: 9781119497110

de Lima Silva, D.F., de Almeida Filho, A.T.: Sorting with TOPSIS through boundary and characteristic profiles. Comput. Ind. Eng. **141**, 106328 (2020). https://doi.org/10.1016/j.cie.2020.106328

de Lima Silva, D.F., Ferreira, L., de Almeida-Filho, A.T.: A new preference disaggregation TOPSIS approach applied to sort corporate bonds based on financial statements and expert's assessment. Expert Syst. Appl. **152**, 113369 (2020). https://doi.org/10.1016/j.eswa.2020.113369

Nemery, P.: An outranking-based sorting method for partially ordered categories. In: DIMACS, Workshop and Meeting of the COST Action ICO602, Université Paris Dauphine, Paris, 28–31 October (2008)

Olson, D.L., Delen, D.: Advanced Data Mining Techniques. Springer, Berlin (2008). ISBN: 9783540769163

Opricovic, S.: Multicriteria Optimization in Civil Engineering Systems. Faculty of Civil Engineering, University of Belgrade, Belgrade (1998)

Ouenniche, J., Pérez-Gladish, B., Bouslah, K.: An out-of-sample framework for TOPSIS-based classifiers with application in bankruptcy prediction. Technol. Forecast. Soc. Chang. **131**, 111–116 (2018). https://doi.org/10.1016/j.techfore.2017.05.034

Ouenniche, J., Bouslah, K., Perez-Gladish, B., Xu, B.: A new VIKOR-based in-sample-out-of-sample classifier with application in bankruptcy prediction. Ann. Oper. Res. **296**, 495–512 (2021). https://doi.org/10.1007/s10479-019-03223-0

Sabokbar, H.F., Hosseini, A., Banaitis, A., Banaitiene, N.: A novel sorting method topsis-sort: an application for Tehran environmental quality evaluation. E&M Econ. Manag. **19**, 87–104 (2016). https://doi.org/10.15240/tul/001/2016-2-006

Vincke, P.: Multicriteria Decision-Aid. Wiley, New York (1992). ISBN: 9780471931843

Wu, D., Olson, D.L.: A TOPSIS data mining demonstration and application to credit scoring. Int. J. Data Warehouse. Min. **2**, 1–10 (2006). https://doi.org/10.4018/jdwm.2006070102

Zardari, N.H., Ahmed, K., Shirazi, S.M., Yusop, Z.B.: Weighting Methods and their Effects on Multi-Criteria Decision Making Model Outcomes in Water Resources Management. Springer, Cham (2015). ISBN: 9783319125855

Zhu, X., Li, J., Wu, D., Wang, H., Liang, C.: Balancing accuracy, complexity and interpretability in consumer credit decision making: a C-TOPSIS classification approach. Knowl.-Based Syst. **52**, 258–267 (2013). https://doi.org/10.1016/j.knosys.2013.08.004

Chapter 7
Rank Reversal in TOPSIS

A rank reversal (RR) phenomenon is a change in the rank ordering of the alternatives when the number of available alternatives changes or other parameters change in exploiting multi-criteria decision making (MCDM) for evaluation alternatives. The issue of RR lies at the heart of many debates in MCDM and traditional decision-making theories. This chapter will investigate the phenomena and properties of RR in the Technique for Order of Preference by Similarity to Ideal Solution (TOPSIS) and reveal some ways to minimize the phenomenon.

7.1 Rank Reversal

In the area of MCDM or decision theory, rank reversal (RR), rank abnormalities, or rank irregularity, is the change in rank ordering of alternatives when an extra alternative is added to or an existing alternative is deleted from the evaluation set. This phenomenon violates our intuitions and is formally named the axiom of independence of irrelevant alternatives (IIA) in multi-attribute utility theory (von Winterfeldt and Edwards 1986). The Analytic Hierarchy Process (AHP) was the first method in MCDM where RR was identified by Belton and Gear (1983), and Salo and Hämäläinen (1997) attributed the reason to AHP's relative measurement mode. A great number of debates have been published, e.g., Saaty (1990a, b, 1991), Saaty and Vargas (1984). Later, Zanakis et al. (1998) utilized simulation to compare eight MCDM methods: ELimination Et Choice Translating REality (ELECTRE), TOPSIS, Multiplicative Exponential Weighting (MEW), Simple Additive Weighting (SAW), and four versions of AHP. In the RR experiments, the performance orders of methods were SAW and MEW (the best), TOPSIS, AHPs, and ELECTRE (the worst). Even though TOPSIS performed at a high standard, it still suffered from the phenomenon. In addition, Wang and Luo (2009) illustrated that other MCDM methods such as the Borda-Kendall method and Data Envelopment Analysis (DEA), including TOPSIS

H.-S. Shih and D. Olson, *TOPSIS and its Extensions: A Distance-Based MCDM Approach*, Studies in Systems, Decision and Control 447,
https://doi.org/10.1007/978-3-031-09577-1_7

and SAW, where dominated to have RR in the examples they presented. Wang and Triantaphyllou (2008) also mentioned RR situations in PROMETHEE as well as in ELECTRE. Observe that the phenomenon can occur with either the removal or the addition of criteria.

We can classify RR into two types: total and partial. In total RR, the original best solution becomes worst, and the original worst solution becomes the best. In partial RR finds the original ranking shave some reversal. Further organizing the RR phenomenon into five types can be seen in Wikipedia, we refer to de Farias Aires and Ferreira's (2019) five types of RR:

(i) The final rank order of alternatives changes if an irrelevant alternative is added to or removed from the problem;
(ii) The indication of the best alternative changes if a non-optimal alternative is replaced by an inferior alternative;
(iii) The transitivity property is violated if an irrelevant alternative is added to or removed from the problem;
(iv) The transitivity property is violated if the initial decision problem is decomposed into sub-problems, the rankings of smaller problems are in conflict with the overall ranking of the alternatives, and
(v) The final rank order of the alternatives changes if a non-discriminating criterion is removed from the problem.

Note that Wikipedia also defined five types of RR with a slightly different depiction. The above types commonly happen in AHP, and researchers have found many RR cases in other MCDM methods. It is possible to specify other types of RR, but the first type would be the most concerning.

Many experts tried to illustrate the RR phenomenon with different examples, but we found that the examples could be misleading in describing RR. The key issue is the characteristics of non-dominated alternatives. In MCDM, non-dominated alternatives or solutions are preferred more than the rest, and the choice of the non-dominated alternatives will depend on DMs' preferences. If we could not initially differentiate the characteristics in the examples first, the proposed solutions would be questionable. Kaliszewski (2018) looked at the rankings of 50 Polish universities and discovered that their best ranking was assigned to one of the non-dominated universities, and that the best one changed depending on the assigned weight set. This discovery raises an important question: Can the varied rankings be categorized as an RR phenomenon? It would be doubtful to say so. Let's present an example to illustrate the ranking changes on non-dominated alternatives via SAW, which is the method least affected by RR.

Example 7.1 A ten alternative evaluated by three criteria.

There are ten alternatives, A_1 to A_{10}, with three benefit criteria, X_1, X_2, and X_3, to be evaluated by SAW as in Table 7.1. The weight set (w_1, w_2, w_3) is associated with the corresponding criterion.

Table 7.1 Performance of ten illustrative alternatives

Alternative	Criteria		
	X_1	X_2	X_3
A_1	1	10	5
A_2	2	9	5
A_3	3	8	5
A_4	4	7	5
A_5	5	6	5
A_6	6	5	5
A_7	7	4	5
A_8	8	3	5
A_9	9	2	5
A_{10}	10	1	5

The above alternatives are obviously non-dominated alternatives, and any one of them is a justifiable choice, no matter what their ranks are. If there is no extra preference provided by DMs, the evaluation could have little meaning. For instance, if we assigned equal weights to the criteria, the ranks of the alternatives are the same, i.e., all ranked number 1. To further illustrate the impact of weight change on the change of ranks, in Table 7.2 we set the weight of criterion X_1 from 0.47 to 0.48 under the condition that the weight of criterion X_3 is fixed at 0.05. Criteria X_3 can be considered as a wash criterion which has no influence on the decision. Their ranks have dramatically changed, which could be noted as total RR. Please look into the ranking results in Table 7.2 as follows.

Table 7.2 Ranks on ten alternatives with five weight sets

		Rank				
Alternative	A_1	1	1	1	10	10
	A_2	2	2	1	9	9
	A_3	3	3	1	8	8
	A_4	4	4	1	7	7
	A_5	5	5	1	6	6
	A_6	6	6	1	5	5
	A_7	7	7	1	4	4
	A_8	8	8	1	3	3
	A_9	9	9	1	2	2
	A_{10}	10	10	1	1	1
Weight set	w_1	0.47	0.4725	0.475	0.4775	0.48
	w_2	0.48	0.4775	0.475	0.4725	0.47
	w_3	0.05	0.05	0.05	0.05	0.05

Observe that a small change in weight w_1 generates very diverse rankings. Other rankings could definitely be generated under different weight sets. Table 7.2 supports the idea that it is better to check non-dominated alternatives before proposing an algorithm for dealing with RR. Many examples in the past have illustrated their algorithms with troublesome examples, e.g., Li (2009), Kong (2011), García-Cascales and Lamata (2012). They could not indicate that the RR phenomenon could result from the effect of non-domination before proposing the treatments. Furthermore, García-Cascales and Lamata (2012) demonstrated the example with all non-dominated alternatives that made their treatment dubious.

Non-domination also makes the rankings sensitive to the changes in other parameters used in the MCDM decision process. For instance, normalization is a possible source of RR. In AHP, one way to deal with the phenomenon is adopting an absolute mode instead of relative or distributive mode (Saaty 1990b). Apart from avoiding the phenomenon of RR, other thinking focuses on the necessity of the method without RR. Millet and Saaty (2000) argued that rank preservation does not always generate a better outcome because of a lack of meaningful results. They claimed that instead of focusing on RR, the focus should be on the method's validity.

7.2 TOPSIS Rank Reversal

The TOPSIS algorithm has provided seven steps for the evaluation (see Chap. 2). Normalization processes have been the focus of RR. In Chap. 3, we have listed eight common methods for normalization. The standard metric used in TOPSIS is Euclidean, taking the square root of the sum of squared distances (the L_2 metric). Sometimes the first power is used, taking the mean of the sum of absolute distances, the L_1 metric, which has been argued as being more robust that is less affected by outliers. The concept appeared in Kong (2011), García-Cascales and Lamata (2012), Senouci et al. (2016), de Farias Aires and Ferreira (2019). Although any positive metric can be used, the limit would be the L_∞ metric, Tchebycheff or Chebyshev, which is based on the maximum difference (here outliers would dominate). These metric rankings are likely to differ. Technically, this is a form of RR, but it is not a problem—it simply reflects different metrics that place different emphasis on outliers.

The second focus concerns the change in the location of PIS and NIS, which cause RR in TOPSIS. The location of PIS and NIS will be changed when an extreme extra alternative is added to or an extreme existing alternative is deleted from the evaluation set. This form of change has an effect on RR because of rescaling the performance values. Many other experts have tried to expand their scales to cover all alternatives, and the simple way is to introduce fictitious extreme alternatives, historical extreme values, or use values assigned by DMs based on their experience. The new alternatives should involve a scale large enough so that the common alternatives are added to the action set and would have no impact on the PIS or NIS. This idea was employed by Ren et al. (2007), Kong (2011), García-Cascales and Lamata (2012), Senouci et al. (2016), de Farias Aires and Ferreira (2019), Yang (2020).

In Fig. 2.2, we illustrated the choice behavior of TOPSIS and illustrated it by indifference curves with fixed values of their relative closeness. It indicates that curves with values higher than 0.5 are convex to the preference origin, i.e., PIS, which stands for the diminishing marginal rate of substitution on the two criteria, whereas curves with values of less than 0.5 are concave. This irregular condition makes TOPSIS selection uncommon (Opricovic and Tzeng 2004). Since the ranking of alternatives is determined by the value of relative closeness, the more the better. Other experts had focused on redesigning the relative closeness or ranking index. Kuo (2017) thought that no weight on TOPSIS's separation measures could limit its applications and contain a flaw. Although his work did not involve RR, he investigated different ranking indices and their effects, including Eq. (3.4) and its variant, Eqs. (3.5) and (3.6), and found weights on the separation measures indeed had an effect on the rankings of alternatives. Note that the three illustrated indices are in linear combination compared to the numerator and denominator form of original relative closeness. We organize the current studies treating RR in TOPSIS in Table 7.3.

From Table 7.3, the two common methods to avert RR are a normalization procedure and using fixed PIS/NIS. de Farias Aires and Ferreira (2018) confirmed that linear normalization, divided by maximum form, produces the lowest RR over other normalization types. However, there is no objective means to adjust or introduce extreme values to anchor a larger PIS and/or a smaller NIS. It is obvious that there is no way to divert the phenomenon of RR in TOPSIS. The deep understanding is still under development.

7.3 Illustrative Examples

In this section, we will choose a real-world application on mobile network interface selection (Senouci et al. 2016), Table 7.4, for illustration. Using a real example might avert fictitious problems arising from man-made examples.

Example 7.2 A mobile network interface selection problem (Senouci et al. 2016).

The decision matrix, with seven alternatives under five criteria, is listed in Table 7.4 (Senouci et al. 2016).

First, we would like to know if different weight sets have an impact on the rankings of alternatives. Based on the characteristics of criteria, we know that alternatives A_4 (on criterion X_1, X_2, and X_3), A_1 (on X_2, X_3, and X_4), A_3 (on X_2), A_6 (on X_5), and A_7 (on X_5) provide the extreme values. These extreme values will promote the corresponding alternatives to rank higher under a greater weight on various criteria as discussed by Kaliszewski (2018). There could be numerous other weight sets, but we only picked eight sets for illustration. The rankings of the seven alternatives are presented in Table 7.5. The lower part of the table indicates the corresponding weight sets for the rankings. Under these limited combinations of the criteria' weights, the alternatives ranked first are A_1, A_4, A_5, and A_7. The rankings of the remaining alternatives indeed

Table 7.3 Current developments dealing with rank reversal in TOPSIS

No	Improvement	Example	Note
1	(1) Reallocate alternatives in the separation plane $S_i^+ - S_i^-$ (2) Calculate the distance of alternatives $C_i = \sqrt{[S_i^+ - \min(S_i^+)] + [S_i^- - \max(S_i^-)]}$ and rank alternatives by R_i in increasing order (3) If the values of two alternatives are equal, use another formula for selection with a small value, i.e., $C_i = S_i^+ - \min(S_i^+)$	(1) 10 × 7 size illustrative decision matrix (2) Existing non-dominated alternatives	(1) Ren et al. (2007) (2) M-TOPSIS (3) Two new ranking indices
2	Only demonstrate RR phenomenon	(1) 4 × 4 size illustrative decision matrix (2) Existing non-dominated alternatives	Wang and Luo (2009)
3	Define relative ratio of alternatives $C_i = \frac{S_i^-}{\max(S_i^-)} - \frac{S_i^+}{\min(S_i^+)}$ and rank alternatives in descending order	(1) 3 × 2 size illustrative decision matrix (2) Existing non-dominated alternatives	(1) Li (2009) (2) Relative ratio method (3) New ranking index
4	(1) Use linear normalization, max–min (2) Select extreme points by DMs	(1) 5 × 2 size illustrative decision matrix (2) Existing non-dominated alternatives	Kong (2011)
5	(1) Add extreme fictitious alternatives, fixed PIS and NIS (2) Use linear normalization, max	(1) 4 × 2 size illustrative decision matrix (2) All non-dominated alternatives	García-Cascales and Lamata (2012)
6	Integrate design of experiment (DoE) into TOPSIS for financial performance evaluation	(1) 5 × 5 size case for demonstration (2) Existing non-dominated alternatives	İç (2014)
7	(1) Establish a range of ideal solutions on criteria and provide a function for normalization (2) Calculate the variation to the normalized reference ideal, unit multiplying the weights on criteria	(1) 5 × 6 size matrix for evaluating candidates (2) Existing non-dominated alternatives	(1) Cables et al. (2016) (2) RIM-reference method

(continued)

Table 7.3 (continued)

No	Improvement	Example	Note
8	(1) Use linear normalization, max–min (2) Provide a dynamic process for keeping the max/min values for sure alternatives' performance within the range	(1) 7 × 5 size matrix for ranking mobile network interface (2) Existing non-dominated alternatives	(1) Senouci et al. (2016) (2) TOPSIS-based dynamic approach
9	(1) Create weighted proximity index u_{ij}, where $u_{ij} = \max(v_j) - v_{ij}$ for benefit criterion and $u_{ij} = v_{ij} - \min(v_j)$ for cost criterion (2) Define overall proximity value of alternatives $C_i = \sum_{j=1}^{n} u_{ij}$ and rank alternatives in decreasing order	(1) Many size matrices for ranking mobile network interface (2) Existing non-dominated alternatives	(1) Mufazzal and Muzakkir (2018) (2) Proximity indexed value (3) New ranking index
10	(1) Use linear normalization, max or max–min (2) Select extreme points by DMs flexibly	(1) 20 × 3 size matrix for selecting students (2) Existing non-dominated alternatives	(1) de Farias Aires and Ferreira (2019) (2) R-TOPSIS
11	(1) Select extreme points by historical information and re-fix PIS and NIS (2) Differentiate cost or benefit criteria for normalization	(1) 4 × 4 size illustrative decision matrix from Wang and Luo (2009) (2) Existing non-dominated alternatives	(1) Yang (2020) (2) NR-TOPSIS
12	Use cumulative Gaussian/normal probability density function for normalization	(1) Six cases with 10 × 5 size data from CloudHarmony dataset (2) Existing non-dominated alternatives	(1) Tiwari and Kumar (2021) (2) G-TOPSIS

provide great deviations. This illustration tells us that the rankings are vulnerable to changes in weights. Consequently, the RR phenomenon could be considered to be a normal phenomenon. Most importantly, RR should be analyzed under a fixed weight set.

Second, we will look at the impact on rankings by deleting an alternative. Senouci et al. (2016) demonstrated the RR phenomenon by deleting alternative A_3, UMTS, and the new rankings, Rank #9, compared to original ranking, Rank #1, are shown in Table 7.6. By applying linear normalization, dividing by their maximum values for benefit criteria, instead of the original vector normalization, its rankings are marked #10. We can see that the operation of linear normalization could not absolutely eliminate the phenomenon, not as we expected. Further experiment employs Manhattan distance

Table 7.4 Decision matrix for mobile network interface selection

Alternative		Criteria				
		Delay (ms)	Jitter (ms)	Loss rate (%)	Throughput (Mbps)	Cost (cent (¢) for MB)
		X_1	X_2	X_3	X_4	X_5
LTE1	A_1	40	20	1	100	10
LTE2	A_2	50	30	2	60	8
UMTS	A_3	35	20	2	1	45
WiMAX1	A_4	30	20	1	50	10
WiMAX2	A_5	35	25	2	30	6
802.11b	A_6	140	75	2	11	1
802.12g	A_7	100	50	3	54	1
Characteristic		Cost	Cost	Cost	Benefit	Cost
Weight		0.05	0.05	0.1	0.3	0.5

Table 7.5 Rank variation by different weight sets

		Rank							
		#1*	#2	#3	#4	#5	#6	#7	#8
Alternative	A_1	1	5	1	1	1	1	4	2
	A_2	3	4	3	3	3	2	2	4
	A_3	7	7	5	5	7	7	7	5
	A_4	4	6	2	2	2	4	3	1
	A_5	5	3	4	4	4	5	1	3
	A_6	6	2	7	7	6	6	6	7
	A_7	2	1	6	6	5	3	5	6
Weights on criteria	X_1	0.05	0.05	0.5	0.05	0.05	0.05	0.35	0.5
	X_2	0.05	0.05	0.05	0.5	0.05	0.05	0.05	0.05
	X_3	0.1	0.1	0.1	0.1	0.5	0.1	0.05	0.35
	X_4	0.3	0.1	0.3	0.3	0.3	0.5	0.05	0.05
	X_5	0.5	0.7	0.05	0.05	0.1	0.3	0.5	0.05

Note '*' original weights from Senouci et al. (2016)

$L_{p=1}$, instead of Euclidean distance $L_{p=2}$, and the corresponding rankings are Ranks #11, #12, and #13 in Table 7.6. Despite the fact that Ranks #1 and #11 are the same as both with Euclidean distance, the rankings of the remaining four are very different. The simple illustration shows us that RR seems unavoidable.

Third, we additionally run tests on adding an alternative. In previous research given Table 7.3, the operation of addition could be classified into two types: performance values influencing the PIS/NIS or not. Normally, the RR phenomenon would not occur if the PIS/NIS were not changed. Nevertheless, the change to PIS/NIS can

Table 7.6 Ranking results by deleting an alternative

		Rank						Note
		#1*	#9	#10	#11	#12	#13	
Alternative	A_1	1(1)	3	3	1(1)	2	3	The digits in the parentheses are the ranks without A_3
	A_2	3(3)	5	4	3(3)	5	5	
	A_3	7(–)	–	–	7(–)	–	–	
	A_4	4(4)	6	5	4(4)	6	4	
	A_5	5(5)	4	6	5(5)	4	6	
	A_6	6(6)	2	2	6(6)	3	2	
	A_7	2(2)	1	1	2(2)	1	1	

Note
1. '*' original ranking, Rank #1, is from Senouci et al. (2016), which is by vector normalization and Euclidean distance $L_{p=2}$
2. Ranks #9 and #10 are by vector normalization and linear normalization, respectively. Both distance measures utilize Euclidean distance
3. Ranks #11 and #12 are by vector normalization, and Rank #13 by linear normalization. All distance measures employ Manhattan distance $L_{p=1}$

produce RR. We display the results of the rankings of both types using the same case in Table 7.7. Rank #1 is the original ranking from Senouci et al. (2016). Rank #14 is obtained by adding an extra alternative A_{F1}, whose values are the averages of the performance of all alternatives. The ranking is the same as #1 and the RR phenomenon does not occur. Rank #15 is obtained by adding an alternative A_{F2}, where the extreme values change the PIS. Alternatives with ranks #1, #14, and #15 are by vector normalization, and ranks #16, #17, and #18 are by linear normalization, which keep the same order as the former corresponding arrangements. The ranks in the parentheses are presented without counting the ranks of the extra alternatives. Rank #14 is the same as Rank #1, but Rank #15 shows a great deal of variety compared to Rank #1. In addition, the rankings with linear normalization are rather consistent, i.e., the rankings of #16, #17, and #18 are the same, and show no RR phenomenon, though these rankings are diverse in contrasting with Rank #1. The observation verifies results given in Table 7.3. Remark that all these rankings are operated by Euclidean distance.

Fourth, many experts claim that adding an extreme alternative could prevent the RR phenomena. We thus add another extreme alternative, A_{F3}, and see the change in the rankings. Their rankings are listed in Table 7.8. To observe the phenomenon, we arrange two settings: Rank #19 considers two extra alternatives, one being extreme alternative A_{F3} and the other is A_{F2} as before, and Rank #20 only counts the extreme alternative A_{F3}. When compared to Rank #1, Ranks #19 and #20 generate different ranks, and RR occurs. However, with the linear normalization, Ranks #16, #21 and #22 preserve the rank, which follow the same order as the former corresponding arrangements. These results confirm the suggestions by Kong (2011), García-Cascales and Lamata (2012), Senouci et al. (2016), de Farias Aires and

Table 7.7 Ranking results by adding alternatives

		Rank						Note
		#1*	#14	#15	#16	#17	#18	
Alternative	A_1	1	1(1)	3(2)	3	3(3)	4(3)	The digits in the parentheses are the ranks without A_{F1} or A_{F2}
	A_2	3	3(3)	4(3)	4	4(4)	5(4)	
	A_3	7	8(7)	8(7)	7	8(7)	8(7)	
	A_4	4	4(4)	6(5)	5	5(5)	6(5)	
	A_5	5	5(5)	5(4)	6	7(6)	7(6)	
	A_6	6	6(6)	7(6)	2	2(2)	2(2)	
	A_7	2	2(2)	2(1)	1	1(1)	1(1)	
	A_{F1}	–	7(–)	–	–	6(–)	–	Performance [61.43, 34.29, 1.86, 43.71, 11.57] by taking an average on the corresponding values
	A_{F2}	–	–	1(–)	–	–	3(–)	Performance [50, 18, 1, 110, 10] by given

Note
1. '*' original ranking is from Senouci et al. (2016)
2. Ranks #1, #14, and #15 are by vector normalization
3. Ranks #16, #17, and #18 are by linear normalization

Table 7.8 Ranking results by adding an extreme alternative

		Rank						Note
		#1*	#19	#20	#16	#21	#22	
Alternative	A_1	1	4(2)	2(1)	3	5(3)	4(3)	The digits in the parentheses are the ranks without A_{F3} and/or A_{F2}
	A_2	3	5(3)	4(3)	4	6(4)	5(4)	
	A_3	7	9(7)	8(7)	7	9(7)	8(7)	
	A_4	4	8(6)	6(5)	5	7(5)	6(5)	
	A_5	5	6(4)	5(4)	6	8(6)	7(6)	
	A_6	6	7(5)	7(6)	2	3(2)	3(2)	
	A_7	2	2(1)	3(2)	1	2(1)	2(1)	
	A_{F1}	–	–	–	–	–	–	Not included here
	A_{F2}	–	3(–)	–	–	4(–)	–	Performance [50, 18, 1, 110, 10] by given
	A_{F3}	–	1(–)	1(–)	–	1(–)	1(–)	Performance [25, 15, 0.8, 120, 0.8] by given

Note
1. '*' original ranking is from Senouci et al. (2016)
2. Ranks #1, #19, and #20 are from vector normalization
3. Ranks #16, #21, and #22 are from linear normalization

Ferreira (2019), which utilize both linear normalization and add an extreme alternative. These illustrations offer a better understanding of TOPSIS ranking behavior. Observe that all the above rankings are calculated by Euclidean distance.

7.4 Theoretical Insights

According to the previous observation, the RR phenomenon would occur if the PIS or NIS changed, i.e., an alternative with maximum or minimum values on criteria, was added or removed. We now concentrate on the formulation of relative closeness, the ranking index, and compare two existing alternatives to examine the possibility of RR under the condition of an extreme alternative added. Here we assume only PIS changes.

Given two alternatives, p and q, the relative closeness of both alternatives are C_p^* and C_q^*, respectively. Their relative closeness can be represented as $C_p^* = \frac{S_p^-}{S_p^+ + S_p^-}$ and $C_q^* = \frac{S_q^-}{S_q^+ + S_q^-}$, accordingly. If alternative p is preferred to alternative q, an inequality will be

$$\frac{S_p^-}{S_p^+ + S_p^-} > \frac{S_q^-}{S_q^+ + S_q^-} \tag{7.1}$$

After moving the terms of denominators to the other sides of the sign, a simple formula is obtained.

$$S_p^- S_q^+ - S_p^+ S_q^- > 0 \tag{7.2}$$

This is the basic formula for judging whether alternative p dominated alternative q. If the inequality is less than zero, alternative q dominated alternative p. In a normal situation, the latter case could be called RR phenomenon, with the variation on some parameters so that the domination is shifted from one to the other.

Let's further depict the inequality in detail. In accordance with the mathematical formulae of TOPSIS, $S_p^- = \sqrt{\sum_{j=1}^{n} (v_{pj} - v_j^-)^2}$, $S_p^+ = \sqrt{\sum_{j=1}^{n} (v_{pj} - v_j^+)^2}$, $S_q^+ = \sqrt{\sum_{j=1}^{n} (v_{qj} - v_j^+)^2}$, and $S_q^- = \sqrt{\sum_{j=1}^{n} (v_{qj} - v_j^-)^2}$, where $j = 1, \ldots, n$ for criteria. The terms v_{ij}, $i = 1, \ldots, m; j = 1, \ldots, n$, are the weighted performance of alternative i on criterion j. The terms of v_j^+ and v_j^- are regarded as the maximum and minimum values of jth benefit criterion, respectively.

Assume that the performance of the jth criterion make a change, and its PIS v_j^+ and NIS v_j^- move to $\overline{v_j^+}$ and $\overline{v_j^-}$ accordingly. The corresponding separation measures and relative closeness for alternative p and q are $\overline{S_p^+}$, $\overline{S_p^-}$, $\overline{C_p^*}$, and $\overline{S_q^+}$, $\overline{S_q^-}$, $\overline{C_q^*}$, respectively. The modified formulation from Eq. (7.2) will be as follows.

$$\overline{S_p^-}\,\overline{S_q^+} - \overline{S_p^+}\,\overline{S_q^-} > 0 \tag{7.3}$$

This equation provides a standard for examining if the RR occurs or not. If Eq. (7.3) holds, the change will not generate RR. However, if $\overline{S_p^-}\,\overline{S_q^+} - \overline{S_p^+}\,\overline{S_q^-} < 0$, RR happens.

Let's illustrate the situation in detail with the original performance measure. The terms x_{ij}, $i = 1,\ldots, m$; $j = 1,\ldots, n$, are the performance of alternative i on criterion j. The terms of x_j^+ and x_j^- are the maximum and minimum values of *jth* benefit criterion, respectively. The movement, due to the change, its extreme values are marked as $\overline{x_j^+}$ and $\overline{x_j^-}$, respectively. However, we only count the change on the maximum value for its simplicity. We use the ratio Δ_j^+ to describe the change on PIS as

$$\Delta_j^+ = \frac{\overline{x_j^+}}{x_j^+} \tag{7.4}$$

where $\Delta_j^+ \geq 1$.

After setting equal weights on criteria, the above equation could be represented as follows.

$$\Delta_j^+ = \frac{\overline{v_j^+}}{v_j^+} \tag{7.5}$$

For any alternative, the new normalized value has a change as

$$\overline{v_{ij}} = \frac{v_{ij}}{\Delta_j^+} \; on\ jth\ criterion \tag{7.6}$$

For any change on criterion l, four new corresponding separation measures, to PIS and NIS, can be represented as follows. Remark that the information of lth criterion is drawn out from the original summation sign and the rest is kept in the original form, the last term.

$$\overline{S_p^-} = \sqrt{(v_{pl}/\Delta_l^+ - v_l^-)^2 - (v_{pl} - v_l^-)^2 + \sum\nolimits_{\substack{j=1 \\ j\neq l}}^{n} (v_{pj} - v_j^-)^2} \tag{7.7}$$

$$\overline{S_q^+} = \sqrt{(v_{ql}/\Delta_l^+ - v_l^-)^2 - (v_{ql} - v_l^-)^2 + \sum\nolimits_{\substack{j=1 \\ j\neq l}}^{n} (v_{qj} - v_j^-)^2} \tag{7.8}$$

$$\overline{S_p^+} = \sqrt{(v_{pl}/\Delta_l^+ - v_l^+)^2 - (v_{pl} - v_l^+)^2 + \sum\nolimits_{\substack{j=1 \\ j\neq l}}^{n} (v_{pj} - v_j^+)^2} \tag{7.9}$$

$$\overline{S_q^-} = \sqrt{(v_{ql}/\Delta_l^+ - v_l^+)^2 - (v_{ql} - v_l^+)^2 + \sum\nolimits_{\substack{j=1 \\ j\neq l}}^{n} (v_{qj} - v_j^+)^2} \tag{7.10}$$

Hence, Eq. (7.2) could be modified as follows.

$$\frac{\sqrt{(v_{pl}/\Delta_l^+ - v_l^-)^2 - (v_{pl} - v_l^-)^2 + \sum_{\substack{j=1\\ j\neq l}}^{n} (v_{pj} - v_j^-)^2}}{\sqrt{(v_{ql}/\Delta_l^+ - v_l^-)^2 - (v_{ql} - v_l^-)^2 + \sum_{\substack{j=1\\ j\neq l}}^{n} (v_{qj} - v_j^-)^2}} > \frac{\sqrt{(v_{pl}/\Delta_l^+ - v_l^+)^2 - (v_{pl} - v_l^+)^2 + \sum_{\substack{j=1\\ j\neq l}}^{n} (v_{pj} - v_j^+)^2}}{\sqrt{(v_{ql}/\Delta_l^+ - v_l^+)^2 - (v_{ql} - v_l^+)^2 + \sum_{\substack{j=1\\ j\neq l}}^{n} (v_{qj} - v_j^+)^2}} \tag{7.11}$$

To illustrate that the change Δ_j^+ has an impact on the Eq. (7.11), we assign a gap ε between the values of both sides.

$$\varepsilon = \frac{\sqrt{(v_{pl}/\Delta_l^+ - v_l^-)^2 - (v_{pl} - v_l^-)^2 + \sum_{\substack{j=1\\ j\neq l}}^{n} (v_{pj} - v_j^-)^2}}{\sqrt{(v_{ql}/\Delta_l^+ - v_l^-)^2 - (v_{ql} - v_l^-)^2 + \sum_{\substack{j=1\\ j\neq l}}^{n} (v_{qj} - v_j^-)^2}} - \frac{\sqrt{(v_{pl}/\Delta_l^+ - v_l^+)^2 - (v_{pl} - v_l^+)^2 + \sum_{\substack{j=1\\ j\neq l}}^{n} (v_{pj} - v_j^+)^2}}{\sqrt{(v_{ql}/\Delta_l^+ - v_l^+)^2 - (v_{ql} - v_l^+)^2 + \sum_{\substack{j=1\\ j\neq l}}^{n} (v_{qj} - v_j^+)^2}} \tag{7.12}$$

It is obvious that if $\varepsilon > 0$, alternative p always dominates alternative q under any change of the associated parameters, there is no RR. However, if $\varepsilon < 0$ for some cases, the RR occurs. This means that alternative q would dominate alternative p in these cases. We thus make use of the ratio of the change in PIS/NIS as an index to evaluate the RR phenomenon. Consequently, the relationship between PIS/NIS change, a value being greater than 1, and the negative value of the gap ε could supply a way to know the RR occurrence. Let's see the illustration in Example 7.3.

Example 7.3 A four alternative example with three criteria.

We modify the fighter selection problem of Hwang and Yoon (1981) by only considering the first three benefit criteria for the interpretation as shown in Table 7.9.

We first execute a specified TOPSIS algorithm, equal weights on criteria and a linear normalization. The ranks are $A_2 \succ A_3 \succ A_4 \succ A_1$. For further explanation, a new alternative A_5 is added with the performance on criteria [2.5, 2,700, 21,000]. Since there is no change in the PIS, the new ranks are $A_5 \succ A_2 \succ A_3 \succ A_4 \succ A_1$. If we remove A_5, the ranking order will remain the same as the original ranks, i.e., no RR. Then we try to increase the performance value of alternative A_5 on criterion X_2 gradually and observe the possibility of RR. Here for a better comparison, let's

Table 7.9 Decision matrix of the four alternatives with three criteria

Alternative	Criteria			Note
	X_1	X_2	X_3	
A_1	2.0	1,500	20,000	All benefit criteria
A_2	2.5	2,700	18,000	
A_3	1.8	2,000	21,000	
A_4	2.2	1,800	20,000	

assign A_3 and A_4 to be A_p and A_q, respectively. It is also noted that A_2 and A_3 are non-dominated alternatives based on the given data.

As long as the change makes PIS shift outward from the original PIS, the ratio of change Δ_j^+ will be greater than 1. For instance, if $\Delta_j^+ = 2$, the performance value of alternative A_5 on criterion X_2 would be 5,400, double the original value. The results of the ratio of change, from 1 to 3, are shown in Table 7.10. If there is a negative value

Table 7.10 Effect of the change ratio Δ_j^+ in PIS on rank reversal

Δ_j^+	$\overline{S_p^-}$	$\overline{S_q^+}$	$\overline{S_p^+}$	$\overline{S_q^-}$	ε	Rank reversal?	Note
1.0	0.077961	0.119154	0.127199	0.072277	0.000096	No	$A_3 \succ A_4$
1.1	0.073598	0.138185	0.143398	0.070611	0.000045	No	$A_3 \succ A_4$
1.2	0.070098	0.154272	0.158069	0.069317	–0.000143	Yes	$A_4 \succ A_3$
1.3	0.067248	0.167998	0.171098	0.068293	–0.000387	Yes	$A_4 \succ A_3$
1.4	0.064897	0.179828	0.182619	0.067469	–0.000651	Yes	$A_4 \succ A_3$
1.5	0.062937	0.190120	0.192819	0.066798	–0.000914	Yes	$A_4 \succ A_3$
1.6	0.061286	0.199150	0.201882	0.066243	–0.001168	Yes	$A_4 \succ A_3$
1.7	0.059884	0.207134	0.209974	0.065779	–0.001408	Yes	$A_4 \succ A_3$
1.8	0.058682	0.214243	0.217231	0.065388	–0.001632	Yes	$A_4 \succ A_3$
1.9	0.057646	0.220612	0.223772	0.065056	–0.001840	Yes	$A_4 \succ A_3$
2.0	0.056747	0.226351	0.229694	0.064770	–0.002033	Yes	$A_4 \succ A_3$
2.1	0.055961	0.231547	0.235078	0.064524	–0.002211	Yes	$A_4 \succ A_3$
2.2	0.055271	0.236275	0.239992	0.055271	–0.002375	Yes	$A_4 \succ A_3$
2.3	0.054661	0.240595	0.244495	0.064121	–0.002526	Yes	$A_4 \succ A_3$
2.4	0.054121	0.244557	0.248635	0.063956	–0.002666	Yes	$A_4 \succ A_3$
2.5	0.053640	0.248204	0.252454	0.063810	–0.002796	Yes	$A_4 \succ A_3$
2.6	0.053209	0.251571	0.255988	0.063680	–0.002915	Yes	$A_4 \succ A_3$
2.7	0.052823	0.254691	0.259266	0.063564	–0.003027	Yes	$A_4 \succ A_3$
2.8	0.052475	0.257588	0.262316	0.063460	–0.003130	Yes	$A_4 \succ A_3$
2.9	0.052160	0.260287	0.265160	0.063367	–0.003226	Yes	$A_4 \succ A_3$
3.0	0.051874	0.262807	0.267818	0.063282	–0.003315	Yes	$A_4 \succ A_3$

on the gap, $\varepsilon < 0$, the RR occurs. This means that alternative p dominates alternative q originally, and q dominates p later. The ratio would be a critical factor in knowing the RR phenomenon. We demonstrate the effect in the following table.

As the change ratio Δ_j^+ increased, $\overline{S_p^-}$ and $\overline{S_q^-}$ decreased, $\overline{S_p^+}$ and $\overline{S_q^+}$ increased as well. It would be difficult to judge if the RR occurs by these four measures. Nonetheless, the gap ε indeed provides an index to perceive the RR phenomenon. In the beginning of the table, $\Delta_j^+ = 1$ and obtains $\varepsilon = 0.000096$. Based on the statement of Eq. (7.11), there is no RR. If the change is small, less than 1.1, on PIS, the gap ε keeps a positive value and no RR happens, e.g., $A_3 \succ A_4$. After the change ratio starts from 1.2, ε value becomes negative, and the RR occurs, $A_4 \succ A_3$. When we make the change larger, the more negative the value of the gap ε will be, and the phenomenon will not revert with larger values. The phenomenon is demonstrated by the smoothly decreasing values. Thus, the change ratio could be employed as an effective index for seeing the phenomenon. Furthermore, being dependent on the change ratio in PIS, the ranks of the existing alternatives should reverse. Many works suggest that introducing an extreme alternative could avoid RR, but our illustration shows that there is no theoretical way to avoid RR.

7.5 Summary and Discussions

This chapter has investigated the RR phenomenon in TOPSIS. Rank changes in evaluation are common in MCDM due to non-domination. Hence, it is better to identify non-dominated alternatives before the evaluation, and to be aware of the possible options with the highest rank. In addition to the knowledge of the alternatives, two proposals from the literature could divert the RR phenomenon when using the TOPSIS algorithm: (i) using linear normalization on criteria performance; (ii) keeping PIS/NIS as large as possible, even introducing fictitious alternatives; and/or (iii) employing Manhattan distance to linearize the choice behavior as discussed in Sect. 2.3. However, these efforts could only minimize the RR phenomenon (Mufazzal and Muzakkir 2018). As addressed by a counter example, linear normalization could not absolutely eliminate the phenomenon. Both effort of linear normalization and keeping an extreme PIS/NIS could not avoid RR neither. There is still a chance that the phenomenon could occur from our theoretical aspect and the experiment. A proposed gap ε can be an effective index to identify the RR phenomenon.

References

Belton, V., Gear, T.: On a shortcoming of Saaty's method of analytic hierarchies. Omega **11**, 228–230 (1983). https://doi.org/10.1016/0305-0483(83)90047-6

Cables, E., Lamata, M.T., Verdegay, J.L.: RIM-reference ideal method in multicriteria decision making. Inf. Sci. **337–338**, 1–10 (2016). https://doi.org/10.1016/j.ins.2015.12.011

de Farias Aires, R.F., Ferreira, L.: The rank reversal problem in multi-criteria decision making: a literature review. Pesqui. Oper. **38**, 331–362 (2018). https://doi.org/10.1590/0101-7438.2018.038.02.0331

de Farias Aires, R.F., Ferreira, L.: A new approach to avoid rank reversal cases in the TOPSIS method. Comput. Ind. Eng. **132**, 84–97 (2019). https://doi.org/10.1016/j.cie.2019.04.023

García-Cascales, M.S., Lamata, M.T.: On rank reversal and TOPSIS method. Math. Comput. Model. **56**, 123–132 (2012). https://doi.org/10.1016/j.mcm.2011.12.022

Hwang, C.L., Yoon, K.: Multiple Attribute Decision Making: A State-of-the-Art Survey. Springer-Verlag, Berlin (1981). ISBN: 9783540105589

İç, Y.T.: A TOPSIS based design of experiment approach to assess company ranking. Appl. Math. Comput. **227**, 630–647 (2014). https://doi.org/10.1016/j.amc.2013.11.043

Kaliszewski, I.: Rankings revisited. Decyzje **2018**(30), 75–84 (2018). https://doi.org/10.7206/DEC.1733-0092.115

Kong, F.: Rank reversal and rank preservation in TOPSIS. Adv. Mater. Res. **204–210**, 36–41 (2011). https://doi.org/10.4028/www.scientific.net/amr.204-210.36

Kuo, T.: A modified TOPSIS with a different ranking index. Eur. J. Oper. Res. **260**, 152–160 (2017). https://doi.org/10.1016/j.ejor.2016.11.052

Li, D.F.: Relative ratio method for multiple attribute decision making problems. Int. J. Inf. Technol. Decis. Mak. **8**, 289–311 (2009). https://doi.org/10.1142/S0219622009003405

Millet, I., Saaty, T.L.: On the relativity of relative measures – accommodating both rank preservation and rank reversals in the AHP. Eur. J. Oper. Res. **121**, 205–212 (2000). https://doi.org/10.1016/S0377-2217(99)00040-5

Mufazzal, S., Muzakkir, S.M.: A new MCDM method based on proximity indexed value for minimizing rank reversals. Comput. Ind. Eng. **119**, 427–438 (2018). https://doi.org/10.1016/j.cie.2018.03.045

Opricovic, S., Tzeng, G.H.: Compromise solution by MCDM methods: a comparative analysis of VIKOR and TOPSIS. Eur. J. Oper. Res. **156**, 445–455 (2004). https://doi.org/10.1016/S0377-2217(03)00020-1

Ren, L., Zhang, Y., Wang, Y., Sun, Z.: Comparative analysis of a novel M-TOPSIS method and TOPSIS. Appl. Math. Res. eXpress **2007**, abm005 (2007). https://doi.org/10.1093/amrx/abm005

Saaty, T.L.: An exposition on the AHP in reply to the paper "remarks on the analytic hierarchy process." Manage. Sci. **36**, 259–268 (1990a). https://doi.org/10.1287/mnsc.36.3.259

Saaty, T.L.: How to make a decision: the analytic hierarchy process. Eur. J. Oper. Res. **48**, 9–26 (1990b). https://doi.org/10.1016/0377-2217(90)90057-I

Saaty, T.L.: Response to Holder's comments on the analytic hierarchy process. J. Oper. Res. Soc. **42**, 909–914 (1991). https://doi.org/10.1057/jors.1991.176

Saaty, T.L., Vargas, L.G.: The legitimacy of rank reversal. Omega **12**, 513–516 (1984). https://doi.org/10.1016/0305-0483(84)90052-5

Salo, A.A., Hämäläinen, R.P.: On the measurement of preferences in the analytic hierarchy process. J. Multi-Criteria Decis. Anal. **6**, 309–319 (1997). https://doi.org/10.1002/(SICI)1099-1360(199711)6:6%3c309::AID-MCDA163%3e3.0.CO;2-2

Senouci, M.A., Mushtaq, M.S., Hoceini, S., Mellouk, A.: TOPSIS-based dynamic approach for mobile network interface selection. Comput. Netw. **107**, 304–314 (2016). https://doi.org/10.1016/j.comnet.2016.04.012

Tiwari, R.K., Kumar, R.: G-TOPSIS: a cloud service selection framework using Gaussian TOPSIS for rank reversal problem. J. Supercomput. **77**, 523–562 (2021). https://doi.org/10.1007/s11227-020-03284-0

von Winterfeldt, D., Edwards, D.: Decision Analysis and Behavioral Research. Cambridge University Press, New York (1986). ISBN: 9780521273046

Wang, X., Triantaphyllou, E.: Ranking irregularities when evaluating alternatives by using some ELECTRE methods. Omega **36**, 45–63 (2008). https://doi.org/10.1016/j.omega.2005.12.003

Wang, Y.M., Luo, Y.: On rank reversal in decision analysis. Math. Comput. Model. **49**, 1221–1229 (2009). https://doi.org/10.1016/j.mcm.2008.06.019

Yang, W.: Ingenious solution for the rank reversal problem of TOPSIS method. Math. Probl. Eng. **2020**, 9676518 (2020). https://doi.org/10.1155/2020/9676518

Zanakis, S.H., Solomon, A., Wishart, N., Dublish, S.: Multi-attribute decision making: a simulation comparison of select methods. Eur. J. Oper. Res. **107**, 507–529 (1998). https://doi.org/10.1016/S0377-2217(97)00147-1

Chapter 8
Other Related Methods

There are a number of methods related to TOPSIS. Although they are not the main focus of the book, we illustrate three of these supplemental TOPSIS methods: fuzzy TOPSIS, VIKOR (VlseKriterijumska Optimizacija I Kompromisno Resenje from Serbian), and BWM (Best Worst Method).

8.1 Fuzzy TOPSIS

After fuzzy sets were introduced by Zadeh (1965), it has been widely applied to decision making models. Negi (1989) was an initial presenter of fuzzy TOPSIS (also see Chen and Hwang 1992), and he has since attracted a large number of followers as the contents of the fuzzy set research area have grown. Palczewski and Saabun (2019) reviewed applications of fuzzy TOPSIS over the period 2009–2018 and categorized methods as (traditional) fuzzy, hierarchical fuzzy, intuitionistic fuzzy, hesitant fuzzy, and type-2 fuzzy. Pythagorean fuzzy TOPSIS was launched by Zhang and Xu (2014) and interval TOPSIS by Tsaur (2011). It is noted that hesitant fuzzy and Pythagorean fuzzy are two major branches of intuitionistic fuzzy methods, and type-2 fuzzy models concern the uncertainty in membership functions (Wikipedia). Although there have been rich advances. typical fuzzy distance measures, PIS/NIS settings, and defuzzification methods have been addressed in Chap. 4. Here we illustrate some basic materials starting from fuzzy number operations.

H.-S. Shih and D. Olson, *TOPSIS and its Extensions: A Distance-Based MCDM Approach*, Studies in Systems, Decision and Control 447,
https://doi.org/10.1007/978-3-031-09577-1_8

8.1.1 Fuzzy Numbers

As the decision space shifted from a crisp domain to a fuzzy domain, the preference structure has changed. In the new structure, operations should be re-defined for fuzzy numbers and their relations. Let's define the fuzzy set first.

Definition 8.1 Fuzzy set (Zadeh 1965; Wikipedia).

A fuzzy set is a pair (U, m) where U is a set and m: $U \rightarrow [0, 1]$ is a membership. The reference set U is called the universe of discourse, and for each $x \in U$, the value $m(x)$ is called the grade of membership of x in (U, m). The function $m = \mu_A$ is called the membership function of the fuzzy set $A = (U, m)$.

In other words, a fuzzy subset is an ordered set of pairs $A = \{(x, \mu_A) | x \in R\}$ where the element x takes its number on the real line R and the degree of membership $\mu_A(x) \in [0, 1]$. We sometimes use $\tilde{A}$ to denote the fuzzy set. If X in R is a collection of objects donated by x, a fuzzy set in X will be

$$\tilde{A} = \{x, \mu_{\tilde{A}}(x) | x \in X\} \tag{8.1}$$

For example, $\tilde{A}$ = {real numbers considerably larger than 10}. With the notation $\tilde{A} = \{x, \mu_{\tilde{A}}(x) | x \in X\}$ where $\mu_{\tilde{A}}(x) = (1 + (x - 10)^{-2})^{-1}$ if $x > 10$; otherwise, $\mu_{\tilde{A}}(x) = 0$ if $x \leq 10$ (Zimmermann 2001).

A fuzzy number is a generalization of a regular, real number within the interval [0, 1], and it is generally a special case of a convex, normalized fuzzy set of the real line. Although fuzzy numbers are widely used to allow rough or imprecise expressions of values, they could be established in innumerable ways. Some common shapes could be

(i) a triangular fuzzy number

In triangular fuzzy numbers, a three point fuzzy number consists of a minimum, most likely, and maximum value $\tilde{A}$ = {min, mode, max}. Please see Definition 8.2.

Definition 8.2 Triangular fuzzy number (Zimmermann 1987)

A fuzzy number $\tilde{A}$ is represented with three points as (a, b, c) and its membership function $\mu_{\tilde{A}}(x)$ is defined as follows.

$$\mu_{\tilde{A}}(x) = \begin{cases} 0, & x < a \\ \dfrac{x - a}{b - a}, & a \leq x \leq b \\ \dfrac{c - x}{c - b}, & b < x \leq c \\ 0, & x > c \end{cases}. \tag{8.2}$$

For instance, the "medium" in Fig. 3.3 could be attributed to this category, its membership $\tilde{A}$ = {0.3, 0.5, 0.7}. The $\mu_{\tilde{A}}(x) = 0$ if $x < 0.3$ and $x > 0.7$; $\mu_{\tilde{A}}(x) = 1$

at $x = 0.5$; $\mu_{\tilde{A}}(x) \in [0, 1)$ is the left-hand-side of the triangular fuzzy number and $\mu_{\tilde{A}}(x) \in (1, 0]$ is the right-hand-side of the triangular fuzzy number.

(ii) a trapezoidal fuzzy number

A trapezoidal fuzzy number consists essentially of a general triangular fuzzy number with a range of full membership at the apex, i.e., with four values. Please see Definition 8.3.

Definition 8.3 Trapezoidal fuzzy number (Buckley and Eslami, 2002).

A fuzzy number $\tilde{A}$ is represented with four points as (a, b, c, d) and its membership function $\mu_{\tilde{A}}(x)$ is defined as follows.

$$\mu_{\tilde{A}}(x) = \begin{cases} 0, \ x < a \\ \frac{x-a}{b-a}, \ a \le x < b \\ 1, \ b \le x \le c \\ \frac{d-x}{d-c}, \ c < x \le d \\ 0, \ x > d \end{cases} \tag{8.3}$$

For example, the "very high" in Fig. 3.3 could be attributed to this category, $\tilde{A} = \{0.8, 0.9, 1.0, 1.0\}$, where $\mu_{\tilde{A}}(x) = 1$ at $x \in [0.9, 1]$ and $\mu_{\tilde{A}}(x) = 0$ at $x > 1$ and $x < 0.8$ as well. In addition, the five linguistic terms are in Fig. 3.3, and the corresponding fuzzy numbers are shown in Table 8.1. It is obvious the left-hand-side and right-hand-side of the slopes could be defined as non-linear functions in Definitions 8.2. and 8.3.

(iii) a *L-R* fuzzy number

Dubois and Prade (1980) defined a special type of fuzzy number representing the left-hand-side (L) and right-hand-side (R) reference functions. The shape functions iff (i) $L(x) = L(-x)$; (ii) $L(0) = 1$; and L is non-decreasing on $[0, +\infty)$. Please see the following definition.

Table 8.1 Examples of fuzzy numbers

Linguistic variable	Fuzzy number	Note
Very low	(0, 0, 0.1, 0.2)	Trapezoidal membership
Low	(0.1, 0.25, 0.4)	Triangular membership
Medium	(0.3, 0.5, 0.7)	Triangular membership
High	(0.6, 0.75, 0.9)	Triangular membership
Very high	(0.8, 0.9, 1.0, 1.0)	Trapezoidal membership

Definition 8.4 *L-R* type fuzzy number (Dubois and Prade 1980).

A fuzzy number $\tilde{A}$ is said to be an *L-R* type fuzzy number iff

$$\mu_{\tilde{A}}(x) = \begin{cases} L\left(\frac{m-x}{\alpha}\right) for\ x \leq m \\ R\left(\frac{x-m}{\beta}\right) for\ x \geq m \end{cases} \tag{8.4}$$

where m is a real number and the mean value of $\tilde{A}$, and α and β are the left and right spreads, respectively.

Zimmermann (2001) provided an example of the *L-R* type fuzzy number, which $L(x) = \frac{1}{1+x^2}$ and $R(x) = \frac{1}{1+2|x|}$ with $m = 5$, $\alpha = 2$ and $\beta = 3$. Then,

$$\mu_5(x) = \begin{cases} L\left(\frac{5-x}{2}\right) = \frac{1}{1+\left(\frac{5-x}{2}\right)^2} for\ x \leq 5 \\ R\left(\frac{x-5}{3}\right) = \frac{1}{1+2\left|\frac{x-5}{3}\right|} for\ x \geq 5 \end{cases} \tag{8.5}$$

8.1.2 Fuzzy Operations

In comparison to the crisp domain, fuzzy operations need to be defined. In the earlier years, Kaufmann and Gupta (1991) presented fuzzy arithmetic with uncertain numbers. Later, numerous set-theoretic operations with extension principle are evolved. Chen and Hwang (1992) accumulated them by fuzzy operations (basics of fuzzy sets, set-theoretic operations, the extension principle and fuzzy arithmetic) and fuzzy ranking methods (using degree of optimality, using Hamming distance, using α-cut, using comparison function, using fuzzy mean and spread, using proportion to the ideal, using left and right scores, with centroid index, using area measurement, and linguistic ranking method). Some later developments could be found in Dubois and Perny (2016). To make a decision in a fuzzy domain, ranking fuzzy numbers is the core process of comparison. Because a couple of methods could produce results that are counterintuitive (Nguyen 2017), a development of a robust method is still being sought.

8.1.3 Fuzzy TOPSIS Approach

As TOPSIS has often been applied to a fuzzy environment, Negi (1989) presented a simplified procedure, compared to the procedure of traditional TOPSIS in Chap. 2, as follows.

Step 1: Create a decision or evaluation matrix ***D***.

For comparative purposes, we add this step, Eq. (2.1). For trapezoidal fuzzy numbers, Definition 8.3, the elements of $\boldsymbol{D}$ are $\tilde{x}_{ij} = (a_{ij}, b_{ij}, c_{ij}, d_{ij})$, $i = 1,\ldots, m$ and $j = 1,\ldots, n$.

Step 2: Construct the normalized decision matrix ***R***.

Matrix $\boldsymbol{D}$ is normalized to matrix $\boldsymbol{R}$ with m alternatives and n criteria. Consider linear normalization for its simplicity. Their maximum value $\tilde{x}_j^+ = \max_i\{\tilde{x}_{ij}\} = (a_j^+, b_j^+, c_j^+, d_j^+)$, and minimum value $\tilde{x}_j^- = \min_i\{\tilde{x}_{ij}\} = (a_j^-, b_j^-, c_j^-, d_j^-)$, $i = 1,\ldots, m$ and $j = 1,\ldots, n$, where $a_j^+ = \max_i(a_{ij})$ and $a_j^- = \min x_i(a_{ij})$ for the criterion j, $j = 1,\ldots,n$. For benefit criteria, the normalized value $\tilde{r}_{ij} = \tilde{x}_{ij}/\tilde{x}_j^+$ and $\tilde{r}_{ij} = \tilde{x}_j^-/\tilde{x}_{ij}$ for cost criteria. Please refer to Table 3.1 for detail.

For trapezoidal fuzzy numbers, the normalized values by the fuzzy division operation $(\div)$ will be the following equation:

$$\tilde{r}_{ij} = \begin{cases} \tilde{x}_{ij}(\div)\tilde{x}_j^+ = \left(\frac{a_{ij}}{d_j^+}, \frac{b_{ij}}{c_j^+}, \frac{c_{ij}}{b_j^+}, \frac{d_{ij}}{a_j^+}\right) \text{ for benefit criteria} \\ \tilde{x}_j^-(\div)\tilde{x}_{ij} = \left(\frac{a_j^-}{d_{ij}}, \frac{b_j^-}{c_{ij}}, \frac{c_j^-}{b_{ij}}, \frac{d_j^-}{a_{ij}}\right) \text{ for cost criteria} \end{cases} \tag{8.6}$$

where $\tilde{r}_{ij}$, $i = 1,\ldots, m$ and $j = 1,\ldots, n$, are the elements of $\boldsymbol{R}$.

Step 3: Construct the weighted normalized decision matrix ***V***.

Traditionally, a set of weights $\boldsymbol{w} = (w_1,\ldots, w_n)$, $w_j > 0$, $j = 1,\ldots, n$, is given to the corresponding criterion X_j, where $j = 1,\ldots, n$. The elements of matrix $\boldsymbol{V}$ can be obtained using $v_{ij} = r_{ij}\, w_j$, $\forall\, i, j$. Let the fuzzy trapezoidal fuzzy numbers of weights be described as $\tilde{w}_j = (\alpha_j, \beta_j, \gamma_j, \delta_j)$, $\forall\, j$. For trapezoidal fuzzy numbers, the fuzzy multiplication operation $(\cdot)$ is illustrated as follows.

$$\tilde{v}_{ij} = \tilde{r}_{ij}(.) = \tilde{w}_j = \begin{cases} \left(\frac{a_{ij}}{d_j^+}\alpha_j, \frac{b_{ij}}{c_j^+}\beta_j, \frac{c_{ij}}{b_j^+}\gamma_j, \frac{d_{ij}}{a_j^+}\delta_j\right) \text{ for benefit criteria} \\ \left(\frac{a_j^-}{d_{ij}}\alpha_j, \frac{b_j^-}{c_{ij}}\beta_j, \frac{c_j^-}{b_{ij}}\gamma_j, \frac{d_j^-}{a_{ij}}\delta_j\right) \text{ for cost criteria} \end{cases} \tag{8.7}$$

where $\tilde{v}_{ij}$, $i = 1,\ldots, m$ and $j = 1,\ldots, n$, are the elements of $\boldsymbol{V}$.

Step 4: Determine the PIS $\boldsymbol{V}^+$ and NIS $\boldsymbol{V}^-$, respectively.

In traditional TOPSIS, the determination of PIS and NIS is rather direct. For fuzzy data, PIS $\boldsymbol{V}^+$ and NIS $\boldsymbol{V}^-$ could be obtained by some ranking procedures. Lee and Li (1988) proposed a ranking method for the comparison of fuzzy numbers, and Negi (1989) extended it to trapezoidal fuzzy numbers. The $\boldsymbol{V}^+$ and $\boldsymbol{V}^-$ are then drawn as the largest generalized mean and the smallest generalized mean, individually. The generalized mean $\tilde{M}(\tilde{v}_{ij})$, $\forall\, i, j$, with triangular fuzzy numbers, is measured by the following equation under uniform distributions:

$$\tilde{M} = (\tilde{v}_{ij}) = \frac{a + b + c}{3} \tag{8.8}$$

For trapezoidal fuzzy numbers, the formula is

$$\tilde{M} = (\tilde{v}_{ij}) \doteq \frac{-a_{ij}^2 - b_{ij}^2 + c_{ij}^2 + d_{ij}^2 - a_{ij}b_{ij} + c_{ij}d_{ij}}{3(-a_{ij} - b_{ij} + c_{ij} + d_{ij})} \tag{8.9}$$

For each criterion $X_j, j = 1,\ldots, n$, we will find V^+ with elements $\tilde{v}_j^+$, the greatest mean on v_{ij}, and V^- with elements $\tilde{v}_j^-$, $\forall\, i, j$, the lowest mean, respectively.

Observe that there are other methods to determine PIS/NIS for fuzzy TOPSIS. Please refer to the contents of Table 5.3.

Step 5: Calculate the separation measure between alternative A_i, S_i^+ and S_i^-, and the ideal and the negative-ideal solutions, PIS/NIS, respectively.

The separation measure or distance between alternative A_i and the PIS/NIS is suggested by the following two equations:

$$S_i^+ = \sum_{j=1}^{n} d_{ij}^+ \ for\ alternative\ A_i,\ i = 1, \ldots, m. \tag{8.10}$$

and

$$S_i^- = \sum_{j=1}^{n} d_{ij}^- \ for\ alternative\ A_i,\ i = 1, \ldots, m. \tag{8.11}$$

The difference between two fuzzy memberships $\mu_{v_{ij}}(x)$ and $\mu_{v_j^+}(x)$ is defined as

$$d_{ij}^+ = 1 - \left\{\sup_x\left[\mu_{v_{ij}}(x) \wedge \mu_{v_j^+}(x)\right]\right\} = 1 - L_{ij}^+, \forall\, i, j \tag{8.12}$$

where L_{ij} is the highest degree of similarity of $\mu_{v_{ij}}(x)$ and $\mu_{v_j^+}(x)$.

In the same fashion, the difference between two fuzzy membership functions, $\mu_{v_{ij}}(x)$ and $\mu_{v_j^-}(x)$, is depicted as

$$d_{ij}^- = 1 - \left\{\sup_x\left[\mu_{v_{ij}}(x) \wedge \mu_{v_j^-}(x)\right]\right\} = 1 - L_{ij}^-, \forall\, i, j \tag{8.13}$$

The above two equations could be simplified by the fuzzy distance, proposed by Kaufmann and Gupta (1991), which considers the left-hand-side and right-hand-side of the fuzzy membership functions. For two fuzzy numbers, $\tilde{A}$ and $\tilde{B}$, the distance between them is represented as follows.

$$d(\tilde{A}, \tilde{B}) = d_L(\tilde{A}, \tilde{B}) + d_R(\tilde{A}, \tilde{B}) \tag{8.14}$$

where the distance is the absolute value between the two fuzzy numbers from both sides.

After the above operations, both d_{ij}^{+} and d_{ij}^{-} are crisp numbers.

Step 6: Calculate the relative closeness C_i^* of alternative A_i, $i = 1,\ldots, m$.

The relative closeness or ranking index of alternative A_i, is defined as $C_i^* = \frac{S_i^-}{S_i^+ + S_i^-}$, $\forall\ i$, Eq. (2.6).

Step 7: Rank the preference order of all alternatives.

A set of alternatives A_i, $i = 1,\ldots, m$, can now be preference ranked according to the descending order of the value of C_i^*.

Let's demonstrate the procedure through an example.

Example 8.1 An energy technology selection problem (Kaya and Kahraman 2011).

This study dealt with an energy technology selection problem in Turkey. Seven technologies were included: conventional (A_1), nuclear (A_2), solar (A_3), wind (A_4), hydraulic (A_5), biomass (A_6), and combined heat and power (CHP) (A_7). These options were evaluated over nine criteria: efficiency (X_1), energy efficiency (X_2), investment cost (X_3), operation and maintenance cost (X_4), NO_x emission (X_5), CO_2 emission (X_6), land use (X_7), social acceptability (X_8), and job creation (X_9). The weight assessment was done using AHP with fuzzy pair-wise comparisons, and the performances of energy technologies were judged by three experts with fuzzy scores. Here the judgment assumed that all criteria are benefits.

To simply demonstrate the fuzzy TOPSIS process, we take the fuzzy normalized decision matrix and criteria weights from Kaya and Kahraman (2011) as in Table 8.2. They provided triangular fuzzy data and with crisp criteria weights after conversion. Thus, Steps 1–3 are skipped. In Step 4, the weighted normalized fuzzy data is treated by Eq. (8.8), and the results with PIS/NIS are shown in Table 8.3. Because the operation on fuzzy AHP for weights has been removed, all crisp data is now available. In Step 5, the separation measures between alternatives and the PIS/NIS are obtained by Manhattandistance, and the relative closeness and rankings on alternatives are then acquired through Steps 6 and 7. These results are illustrated in Table 8.4. The rankings are $A_4 \succ A_6 \succ A_3 \succ A_7 \succ A_2 \succ A_5 \succ A_1$, which are closely consistent with the results of Kaya and Kahraman (2011) with only ranks of A_2 and A_5 being switched.

Fuzzy sets could represent linguistic input, but the fuzziness could also arise for other reasons. There are number of evolving types of fuzzy approaches that have been proposed, especially for the domains of type-2 fuzzy, intuitionistic fuzzy, Pythagorean fuzzy and hesitant fuzzy. Whatever fuzzy operations are used, the related fuzzy TOPSIS will work for selection or for sorting problems. Fuzzy group TOPSIS is another branch that has drawn a great deal of attention.

Table 8.2 Fuzzy normalized evaluation with criteria weight for seven technology options (Kaya and Kahraman, 2011)

Alternative	Criteria								
	X_1	X_2	X_3	X_4	X_5	X_6	X_7	X_8	X_9
A_1	(0.084,0.97, 1)	(0.7,0.87,0.97)	(0.57,0.77,0.94)	(0.44,0.64,0.84)	(0.04,0.1,0.24)	(0.07,0.2,0.37)	(0,0.07,0.24)	(0.07,0.24,0.44)	(0.57.0.77,0.94)
A_2	(0.084,0.97, 1)	(0.7,0.84,0.9)	(0.04,0.1,0.24)	(0.9, 1, 1)	(0.07,0.24,0.44)	(0.1,0.3,0.5)	(0.07,0.2,0.37)	(0.07,0.24,0.44)	(0.7,0.9, 1)
A_3	(0.24,0.44,0.64)	(0.3,0.5,0.7)	(0.37,0.57,0.77)	(0.3,0.5,0.7)	(0.84,0.97, 1)	(0.7,0.9, 1)	(0.77,0.94, 1)	(0.7,0.9, 1)	(0.37,0.57,0.77)
A_4	(0.14,0.3,0.5)	(0.5,0.7,0.9)	(0.7,0.9, 1)	(0.77,0.94, 1)	(0.77,0.94, 1)	(0.9, 1, 1)	(0.77,0.94, 1)	(0.9, 1, 1)	(0.3,0.5,0.7)
A_5	(0.5,0.7,0.87)	(0.7,0.9, 1)	(0.5,0.7,0.9)	(0.3,0.5,0.7)	(0.1,0.3,0.5)	(0,0.1,0.3)	(0.1,0.3,0.5)	(0.3,0.5,0.7)	(0.64,0.84,0.97)
A_6	(0.3,0.5,0.7)	(0.44,0.64,0.84)	(0.37,0.57,0.77)	(0.24,0.44,0.64)	(0.7,0.9, 1)	(0.7,0.9, 1)	(0.5,0.7,0.9)	(0.7,0.9, 1)	(0.64,0.84,0.97)
A_7	(0.44,0.64,0.84)	(0.37,0.57,0.77)	(0.37,0.57,0.77)	(0.17,0.37,0.57)	(0.3,0.5,0.7)	(0.3,0.5,0.7)	(0.57,0.77,0.94)	(0.64,0.84,0.97)	(0.5,0.7,0.9)
Weight on criterion	0.085	0.1	0.099	0.113	0.127	0.152	0.108	0.093	0.122

Table 8.3 PIS/NIS for seven technology options

Alternative	Criteria									Note
	X_1	X_2	X_3	X_4	X_5	X_6	X_7	X_8	X_9	
A_1	0.0796	0.0847	0.0752	0.0723	0.0161	0.0324	0.0112	0.0233	0.0927	
A_2	0.0796	0.0813	0.0125	0.1092	0.0318	0.0456	0.0230	0.0233	0.1057	
A_3	0.0374	0.05	0.0564	0.0565	0.1190	0.1317	0.0976	0.0806	0.0695	
A_4	0.0266	0.07	0.0858	0.1021	0.1147	0.1469	0.0976	0.0899	0.0610	
A_5	0.0587	0.0867	0.0693	0.0565	0.0381	0.0203	0.0324	0.0465	0.0996	
A_6	0.0425	0.0640	0.0564	0.0497	0.1101	0.1317	0.0756	0.0806	0.0996	
A_7	0.0544	0.0570	0.0564	0.0418	0.0635	0.0760	0.0821	0.0760	0.0854	
PIS	0.0796	0.0867	0.0858	0.1092	0.1190	0.1469	0.0976	0.0899	0.1057	
NIS	0.0266	0.05	0.0125	0.0418	0.0161	0.0203	0.0112	0.0233	0.0610	

Table 8.4 Fuzzy evaluation results of seven technology options

Alternative	S_i^+	S_i^-	C_i^*	Rank	Note
A_1	0.4329	0.2247	0.3417	7	Conventional
A_2	0.4083	0.2494	0.3792	5	Nuclear
A_3	0.2217	0.4360	0.6629	3	Solar
A_4	0.1258	0.5319	0.8088	1	Wind
A_5	0.4124	0.2453	0.3729	6	Hydraulic
A_6	0.2101	0.4475	0.6805	2	Biomass
A_7	0.3278	0.3298	0.5015	4	Combined heat and power (CHP)

8.1.4 Choquet Integral for TOPSIS

The Choquet integral is a sub-additive or super-additive integral created by Choquet (1953). It was first utilized in statistical mechanics and potential theory, but it made its way into decision theory in 1980, where it is used to calculate the expected utility of an uncertain event. It is applied specifically to membership functions and capacities, and a fuzzy set is involved (Wikipedia). Later, Sugeno (1974) proposed a similar concept, fuzzy measure and fuzzy integral, which are both non-additive measures and suitable for expressing interactions between criteria in multi-criteria formulation and introduced some applications of fuzzy integrals (Grabisch 1996). Tzeng and his colleagues employed this technique in many applications, e.g., Chen and Tzeng (2001), Tzeng et al. (2005a), and Yang et al. (2008). Krishnan et al. (2015) published a short survey on its usage in MCDM, in which seven fuzzy measure identification methods were supplied. In Sect. 3.8.3, we discussed other approaches to handling dependent criteria. Because of the fuzzy operations, we've moved this issue here for easier illustration.

The utilization of Choquet or fuzzy integral for TOPSIS is straightforward, but the contents of the publications are much more complicated. In group decision making, DMs' subjective preferences are usually nonlinear, and thus not independent. Fuzzy measures have been introduced to provide tools not limited by the requirement for independence. A key step in fuzzy group models is the aggregation of DMs' opinions. Tan (2011) presented a generalized interval-valued intuitionistic fuzzy geometric aggregation operator to aggregate individual DM's opinions as part of a TOPSIS model using Choquet integral-based Hamming distance.

The steps in applying Choquet integrals within TOPSIS are as follows (Tan 2011).

Step 1: Obtain decision matrices from the DMs containing interval-valued intuitionistic fuzzy values.

There is the degree of uncertainty that each expert has about each alternative satisfying each criterion, as well as the uncertain degree that each expert considers that each alternative will not satisfy the criteria.

Step 2: Calculate the fuzzy density for each expert.

Step 3: Order the options and apply the interval-valued intuitionistic fuzzy Choquet integral operator.

Step 4: Identify the fuzzy positive-ideal and fuzzy negative-ideal solutions.

Step 5: Confirm the fuzzy density of each criterion.

Step 6: Use the Choquet integral-based Hamming distance to calculate the distance between each alternative and the positive-ideal and negative-ideal solutions.

Step 7: Calculate relative closeness with TOPSIS.

Step 8: Rank the preference order of all alternatives.

The ranking process is the same as that of the traditional TOPSIS algorithm.

The key difference is including Choquet operators in separation measures. Tan (2011) used the fuzzy definition consisting of two intervals, each consisting of elements between 0 and 1, the first interval reflecting the degree of belongingness [*a*, *b*], and the second the degree of non-belongingness [*c*, *d*]. The wider the range, the higher the degree of uncertainty. The ideal interval, or PIS, for belongingness would be [1, 1], while the nadir, or NIS, would be [0, 0]. The Hamming distance counts the distance between two intervals $x = ([a_i, b_i], [c_i, d_i])$ and $y = ([a_i', b_i'], [c_i', d_i'])$, $\forall\, i$, as follows.

$$d(x, y) = \frac{1}{4}\sum_{i=1}^{n} \left|a_i - a_i'\right| + \left|b_i - b_i'\right| + \left|c_i - c_i'\right| + \left|d_i - d_i'\right| \tag{8.15}$$

The weighted Hamming distance multiplies the weight vector times each of the four Hamming distance elements. A score function was defined to compare intuitionistic fuzzy values.

$$-1 \leq Score(x) = \frac{a - c + b - d}{2} \leq 1 \tag{8.16}$$

An accuracy function was also defined.

$$0 \leq Accuracy(x) = \frac{a + c + b + d}{2} \leq 1 \tag{8.17}$$

The *Score*(x) is the difference between the membership and non-membership functions, while *Accuracy*(x) is the sum of the membership and non-membership functions. Order relationships were defined as:

$$\text{If } Score(x) < Score(y), \text{ then } x \text{ is less than } y; \tag{8.18}$$

If Score(x) = Score(y),

If $Accuracy(x) < Accuracy(y)$, then x is less than y;

$$\text{If } Accuracy(x) = Accuracy(y), \text{ then } x \text{ and } y \text{ represent the same information.} \tag{8.19}$$

We demonstrate the process with an example.

Example 8.2 An illustrated example with two DMs.

Assume two DMs, DM#1 and DM#2, selecting between two investments (A_1 and A_2), considering the three criteria of risk (X_1), growth (X_2), and environmental impact (X_3). The fuzzy decision matrix, with higher belongingess entries reflecting favorable views, is shown in Table 8.5.

Here, DM#1 considers A_1 to be more risky, but has better growth and environmental impact likelihoods. DM#2 shares those views but at levels with less certainty. Table 8.6 gives scores and accuracies for the example as follows. These scores indicate that for both DM#1 and DM#2, A_2 is better than A_1 on risk, but worse on both growth and environmental impact. The accuracy measures are high for both, with a little less confidence by DM#1 on growth for A_2 (as well as a bit for DM#2).

Table 8.5 Fuzzy Inputs (belongingness and non-belongingness)

Decision maker	Alternative	Criteria		
		X_1	X_2	X_3
DM#1	A_1	[0.3,0.5], [0.4,0.7]	[0.4,0.6], [0.3,0.5]	[0.6,0.7], [0.2,0.3]
	A_2	[0.6,0.7], [0.2,0.4]	[0.2,0.3], [0.4,0.6]	[0.3,0.4], [0.5,0.7]
DM#2	A_1	[0.4,0.6], [0.4,0.5]	[0.5,0.6], [0.3,0.4]	[0.5,0.6], [0.2,0.4]
	A_2	[0.5,0.7], [0.3,0.4]	[0.4,0.5], [0.3,0.4]	[0.4,0.5], [0.4,0.6]

Table 8.6 Results of the scores and the accuracies of two DMs

Decision maker	Alternative	Score			Accuracy		
		X_1	X_2	X_3	X_1	X_2	X_3
DM#1	A_1	−0.15	0.1	0.4	0.95	0.9	0.9
	A_2	0.35	−0.25	−0.25	0.95	0.75	0.95
DM#2	A_1	0.05	0.2	0.25	0.95	0.9	0.85
	A_2	0.25	0.1	-0.05	0.95	0.8	0.95

8.2 VIKOR

VIKOR (VlseKriterijumska Optimizacija I Kompromisno Resenje, VIKOR [from Serbian]; All/Multiple Criteria Optimization and Compromise Solution) is a multi-criteria decision-making method proposed by Opricovic (1998) based on distance ranking from the ideal solution. For m alternatives $(A_1,\ldots, A_m)$ are evaluated by n criteria $(X_1,\ldots, X_n)$, with their performances x_{ij}, where $i = 1,\ldots, m$ and $j = 1,\ldots, n$. The distance of any alternative to its ideal solution can be represented by the Minkowski L_p metrics as

$$L_p = \left\{ \sum_{j=1}^{n} \left[\frac{w_j \left| x_j^+ - x_{ij} \right|}{\left| x_j^+ - x_j^- \right|} \right]^p \right\}^{1/p} \tag{8.20}$$

where $1 \leq p \leq \infty$, $x_j^+ = \max_i(x_{ij})$ and $x_j^- = \min_i(x_{ij})$ are the maximum and minimum values of jth benefit criterion, respectively, and w_j is the weight of the criterion. For cost criteria, the contents of x_j^+ and x_j^- are switched. Moreover, the denominator term $\left| x_j^+ - x_j^- \right|$ is a scaling factor to normalize the distance for each criterion. Please refer to the traditional TOPSIS algorithm and Tables 3.1 and 3.8.

By using the representation of normalized distance, VIKOR considers the compromise index Q_i, between the average gap S_i and the maximal gap for improvement priority R_i, for alternative i, $i = 1,\ldots, m$, in a multiple dimension space. The corresponding formulae are defined:

$$S_i = \sum_{j=1}^{m} \frac{w_j \left| x_j^+ - x_{ij} \right|}{\left| x_j^+ - x_j^- \right|}, \ \forall j \tag{8.21}$$

$$R_i = \max_j \left\{ \frac{\left| x_j^+ - x_{ij} \right|}{\left| x_j^+ - x_j^- \right|} \right\}, \ \forall j \tag{8.22}$$

$$Q_i = v \frac{S_i - S^+}{S^- - S^+} + (1 - v) \frac{R_i - R^+}{R^- - R^+}, \ \forall j \tag{8.23}$$

where $S^+ = \min_i(S_i)$, $S^- = \max_i(S_i)$, $R^+ = \min_i(R_i)$, and $R^- = \max_i(R_i)$, $\forall\ i$; v, $0 < v < 1$, is the coefficient for the convex combination of S_i and R_i.

Observe that Eqs. (8.21) and (8.22) are the Manhattan and the Tchebycheff distances, respectively. Hence, Eq. (8.23) tries to balance these two distances. Opricovic and Tzeng (2004) claimed that the solution acquired by $\min_i(S_i)$ is with a maximum group utility and the solution earned by $\min_i(R_i)$ is with a minimum individual regret of all alternatives. The former is to minimize the weighted distances to PIS among alternatives or to find the closest alternative to PIS, and the later is considered as the Savage regret among alternatives (Taha 2017). The rankings of alternatives are decided by the values of these three indices. Furthermore, the coefficient v generates a convex combination of indices of S_i and R_i. If $v > 0.5$, the decision mainly relies on the average gap S_i, $\forall\ i$, which considers the voting by majority rule; if $v < 0.5$, the decision concentrates on the maximal gap for improvement priority R_i, $\forall\ i$, which counts decision with veto; if $v \approx 0.5$, a balance between above two indices, the decision is made by consensus. Let's list the procedure of VIKOR as follows (Opricovic and Tzeng 2004).

Step 1: Establish a decision or evaluation matrix ***D***

The matrix consists of m alternatives ($A_1,\ldots, A_m$) and n criteria ($X_1,\ldots, X_n$), with its performance x_{ij}, for $i = 1,\ldots,m$ and $j = 1,\ldots,n$.

Step 2: Determine the best x_j^+ and the worst x_j^- values, $\forall\ j$.

For benefit criteria, $x_j^+ = \max_i(x_{ij})$ and $x_j^- = \min_i(x_{ij})$; $x_j^+ = \min_i(x_{ij})$ and $x_j^- = \max_i(x_{ij})$ for cost criteria.

Step 3: Compute the values of the average gap S_i and the maximal gap for improvement priority R_i, for alternative i, $i = 1,\ldots,m$.

Please refer to Eqs. (8.21) and (8.22)

Step 4: Calculate the compromise index Q_i, $\forall\ i$.

Please consult to Eq. (8.23).

Step 5: Rank the alternatives, sorting by the values S_i, R_i, and Q_i, $i = 1,\ldots,m$, in ascending order.

There are three indices in the list. In general, the lower the value is, the better the alternative performs.

Step 6: Propose as a compromise solution the alternative A_i, which is ranked the best by the measure Q_i, $\forall\ i$, if the following two conditions are satisfied.

Condition 1: Acceptable advantage.

$$Q(A_{i'}) - Q(A_i) \geq \Delta Q \tag{8.24}$$

where $A_{i'}$, $i' = 1,\ldots, m$, is the alternative with the second position in the ranking list by Q, and $\Delta Q = 1/(m-1)$, where m is the total number of the alternatives.

Condition 2: Acceptable stability in decision making.

Alternative A_i must also be the better ranked by the values S_i and R_i, $\forall\ i$. Then, this compromise solution is considered stable within a decision making process.

If one of the above conditions is unsatisfied, a set of compromise solutions is suggested based on: (i) comparison of alternatives $A_{i'}$ and A_i, i and $i' = 1,\ldots, m$, only Condition 2 is unsatisfied, or (ii) alternatives $A_1,\ldots,A_m$, $i = 1,\ldots, m$, if Condition 1 is unsatisfied; and A_m is determined by the relation $Q(A_i)-Q(A_{i'}) < \Delta Q$, opposition to Eq. (8.19), for maximum m.

After ranking the alternatives, Opricovic and Tzeng (2004) recommended an extra step for checking weight stability intervals. The results can provide an image of stable weight intervals for the decision.

It is observed that the gap $\Delta Q = 1/(m-1)$ is implicitly assumed to be uniformly distributed of the alternative performances within the interval [0, 1] for their normalized values. If the performance measures are spread over a small range, as the effect of vector normalization, the assumption would be too loose to make a judgment (Shih 2008). This means that the process will generate many tied alternatives, from which it will be difficult to make a choice. Moreover, it seems that the above two conditions intend to guarantee the selected alternative as preferred over the rest of the alternatives. Although the conditions could be utilized for the alternatives with lower ranks, there would be less meaning for the DMs. After over two decades' development, VIKOR has also been found suffer from rank reversal (Ceballos et al. 2018). Over time, various studies have been proposed, including fuzzy VIKOR and group VIKOR. Interested readers could check a couple of review papers such as Yazdani and Graeml (2014), Gul et al. (2016), Mardani et al. (2016), and Ploskas and Papathanasiou (2019).

We use an example to demonstrate the technique.

Example 8.3 An alternative-fuel bus selection problem (Tzeng et al. 2005b).

The Taiwanese transportation bureau executed a study to select suitable alternative-fuel buses for public transportation in urban areas. Twelve types of alternative-fuel buses were evaluated as: conventional diesel engine (A_1), compressed natural gas, CNG (A_2), liquid propane gas, LPG (A_3), fuel cell (hydrogen) (A_4), methanol (A_5), electric vehicle—opportunity charging (A_6), direct electric charging (A_7), electric bus with exchangeable batteries (A_8), hybrid electric bus with gasoline engine (A_9), hybrid electric bus with diesel engine (A_{10}), hybrid electric bus with CNG engine (A_{11}), and hybrid electric bus with LPG engine (A_{12}). These 12 types of buses were evaluated by 11 criteria: energy supply (X_1), energy efficiency (X_2), air pollution (X_3), noise pollution (X_4), industrial relationship (X_5), costs of implementation (X_6), costs of maintenance (X_7), vehicle capability (X_8), road facility (X_9), speed of traffic flow (X_{10}), and sense of comfort (X_{11}). The normalized performances of the 12 types of buses under 11 criteria and the corresponding averaging weights are listed in Table 8.7 (Tzeng et al. 2005b). In addition, all values are associated with benefit criteria.

Table 8.7 Normalized performance of alternative-fuel buses with corresponding weights

Alternative	Criteria										
	X_1	X_2	X_3	X_4	X_5	X_6	X_7	X_8	X_9	X_{10}	X_{11}
A_1	0.82	0.59	0.18	0.42	0.58	0.36	0.49	0.79	0.81	0.82	0.56
A_2	0.77	0.70	0.73	0.55	0.55	0.52	0.53	0.73	0.78	0.66	0.67
A_3	0.79	0.70	0.73	0.55	0.55	0.52	0.53	0.73	0.78	0.66	0.67
A_4	0.36	0.63	0.86	0.58	0.51	0.59	0.74	0.56	0.63	0.53	0.70
A_5	0.40	0.54	0.69	0.58	0.51	0.52	0.68	0.52	0.63	0.60	0.70
A_6	0.69	0.76	0.89	0.60	0.72	0.80	0.72	0.54	0.35	0.79	0.73
A_7	0.77	0.79	0.89	0.59	0.73	0.80	0.72	0.47	0.44	0.87	0.75
A_8	0.77	0.79	0.89	0.59	0.73	0.80	0.72	0.51	0.48	0.87	0.75
A_9	0.77	0.63	0.63	0.52	0.66	0.63	0.65	0.67	0.70	0.80	0.74
A_{10}	0.77	0.63	0.51	0.58	0.66	0.63	0.65	0.67	0.70	0.80	0.74
A_{11}	0.77	0.73	0.80	0.48	0.63	0.66	0.65	0.67	0.71	0.62	0.78
A_{12}	0.77	0.73	0.80	0.48	0.63	0.66	0.65	0.67	0.71	0.62	0.78
Criteria weights	0.0313	0.0938	0.1661	0.0554	0.0629	0.0829	0.0276	0.1239	0.0805	0.1992	0.0761

Table 8.8 Ranks of alternatives by the values of Q_i with S_i and R_i, $\forall\, i$

Rank by Q_i	Alternative	Q_i	S_i	Rank by S_i	R_i	Rank by R_i	Note
1	A_9	0.1683	0.3315	4	0.0608	1	Hybrid electric bus with gasoline engine
2	A_8	0.1716	0.1852	1	0.1084	4	Electric bus with exchangeable batteries
3	A_6	0.2235	0.2667	3	0.0968	3	Electric vehicle—Opportunity charging
4	A_7	0.2534	0.2077	2	0.1239	7	direct electric charging
5	A_{10}	0.2806	0.3411	5	0.0889	2	HYBRID electric bus with diesel engine
6	A_3	0.4787	0.4058	8	0.1232	5	Liquid propane gas, LPG
7	A_2	0.4802	0.4071	9	0.1232	5	Compressed natural gas, CNG
8	A_{11}	0.5100	0.3594	6	0.1466	8	Hybrid electric bus with CNG engine
8	A_{12}	0.5100	0.3594	6	0.1466	8	Hybrid electric bus with LPG engine
10	A_1	0.8058	0.5554	11	0.1661	11	Conventional diesel engine
11	A_5	0.8519	0.6197	12	0.1584	10	Methanol
12	A_4	0.9251	0.5547	10	0.1994	12	Fuel cell (hydrogen)

After the decision information is organized in Step 1, proceed through Steps 2–5 as shown in Table 8.8 after sorting by the values of Q_i, $\forall\, i$. Step 6 is to find the compromise rankings of these 12 alternatives through checking two conditions. The detail contents are depicted in Table 8.9. The final rankings are: $A_9 \approx A_8 \approx A_6 \approx A_7 \succ A_{10} \succ A_3 \succ A_2 \succ A_{11} \approx A_{12} \succ A_1 \succ A_5 \succ A_4$. TOPSIS rankings are presented in the rightmost column for reference. Owing to the first four alternatives being ties, Pearson's correlation coefficient is only 0.7085, which is considered a little low correlation.

8.3 BWM

Rezaei (2015, 2016) presented the best-worst method (BWM) for multi-criteria decision-making problems. The BWM is based on a systematic pair-wise comparison of decision criteria. After the DM has identified the decision criteria, the DM selects the best criterion and the worst criterion as references. The best criterion plays the most significant function in the decision-making process, whereas the worst criterion

Table 8.9 Compromise ranking with two conditions

Rank by Q_i	Alternative	Acceptable advantage	Acceptable stability			Compromise rank	TOPSIS rank
			Check S_i	Check R_i	OR operation		
1	A_9	–	–	–	–	1	6
2	A_8	Fail	Fail	OK	OK	1	1
3	A_6	Fail	Fail	OK	OK	1	3
4	A_7	Fail	Fail	OK	OK	1	2
5	A_{10}	OK	OK	OK	OK	5	10
6	A_3	OK	OK	OK	OK	6	7
7	A_2	OK	OK	OK	OK	7	8
8	A_{11}	OK	OK	OK	OK	8	4
8	A_{12}	OK	OK	OK	OK	8	4
10	A_1	OK	OK	OK	OK	10	12
11	A_5	OK	OK	OK	OK	11	11
12	A_4	OK	OK	OK	OK	12	9

Note The rank 1st alternative is compared to the rest alternatives

plays the opposite role. Using a number from a predetermined scale, i.e., 1–9, the DM then expresses his or her preference for the best criterion over all other criteria, as well as his or her preference for all criteria over the worst criterion. These two sets of pair-wise comparisons are fed into an optimization problem, with the weights of the criterion selected as the best results (Wikipedia). Even though the best and the worst criteria could be thought of as PIS and NISin TOPSIS, it is closely related to AHP, one of the most popular multi-criteria methods. BWM is fairly recent, yet it has received a great deal of use in multi-criteria applications (Mi et al. 2019). One criticism of AHP is that it places excessive burden on DMs due to the exponential growth in the number of pair-wise comparisons as the complexity of the criteria structure grows. BWM has a simplified structure greatly reducing the number of required pair-wise comparisons. This has the residual effect of reducing inconsistency, although the validity of this as a reason to select BWM is dubious in that more comparisons probably provide additional checks of relative weights. However, it is most likely that as the number of pair-wise comparisons grows, DM's accuracy in inputs will probably decline. It is worth noting that the BWM generates the weight set of criteria using a subjective weighting method, although Rezaie (2015 and 2016) claimed it was a multi-criteria method in the beginning of the development. Hence, the method only plays one part in the decision-making process.

We present the BWM algorithm following Rezaei (2015):

Step 1: The DM creates a set of decision criteria $(X_1,\ldots, X_n), j = 1,\ldots, n$.

Step 2: The DM selects the best and the worst criteria, B and W, respectively.

The choice should be from the set of decision criteria.

Step 3: Compare the preference of the best criterion over all the other criteria.

Using a number between 1 and 9 expresses the preference intensity scale and the resulting best-to-other (BO) vector would be

$$A_B = (a_{B1}, a_{B2}, \ldots, a_{Bn}) \tag{8.25}$$

where a_{Bj}, $j = 1,\ldots, n$, denotes the preference intensity of the best criterion B over criterion j. It is obvious that $a_{BB} = 1$, as the nine-point intensity scale of importance of AHP (Saaty 1980).

Step 4: Compare the preferences of all the criteria over the worst criterion.

Use a number between 1 and 9 to express a preference intensity scale, yielding the resulting other-to-worst (OW) vector is as follows.

$$A_W = (a_{1W}, a_{2W}, \ldots, a_{nW})^{\mathrm{T}} \tag{8.26}$$

where a_{jW}, $\forall j$, indicates the preference intensity of criterion j over the worst criterion W. It is apparent that $a_{WW} = 1$.

Step 5: Using some program, determine the optimal weight set $W^* = (w_1^*, w_2^*, \ldots, w_n^*)$.

To acquire the optimal weights of the criteria, the maximum absolute differences $\left|\frac{w_B}{w_j} - a_{Bj}\right|$ and $\left|\frac{w_j}{w_W} - a_{jW}\right|$ for all j, $\forall j$, are minimized. The concept can be transformed into the following minimax programming model:

$$Min \max_i \{\left|\frac{w_B}{w_j} - a_{Bj}\right|, \left|\frac{w_j}{w_W} - a_{jW}\right|\}$$

s.t.

$$\sum_{j=1}^{n} w_j, \text{ where } w_j > 0, \ \forall j \tag{8.27}$$

For simplicity, the above model can be transformed into a linear program by introducing $2j$ additional constraint sets. Observe that these additional constraints can be considered implicitly by treating them as parametric upper bounds (Ahuja 1985).

$$
\begin{aligned}
& Min\, \xi \\
& \text{s.t.} \\
& \left| \frac{w_B}{w_j} - a_{Bj} \right| \le \xi, \ \forall\ j \\
& \left| \frac{w_j}{w_W} - a_{jW} \right| \le \xi, \ \forall\ j \\
& \sum_{j=1}^{n} w_j = 1, \ \text{where } w_j > 0, \ \forall\ j
\end{aligned} \tag{8.28}
$$

Rezaei stated that since pairwise comparisons are not always fully consistent, the above model can yield multiple optimal solutions. He gave a procedure to identify the range of weight values implied by pair-wise comparison input. The procedure is to run the original model to identify the optimal epsilon, then optimize (both maximize and minimize in turn) each of the weights subject to the original constraints plus setting epsilon equal to the optimal identified (rounding up will avoid infeasible solutions). Later, Rezaei (2016) suggested a linear MWM model for obtaining a single optimal solution as follows.

$$
\begin{aligned}
& Min\, \xi^L \\
& \text{s.t.} \\
& \left| w_B - a_{Bj} w_j \right| \le \xi^L, \ \forall\ j \\
& \left| w_j - a_{jW} w_W \right| \le \xi^L, \ \forall\ j \\
& \sum_{j=1}^{n} w_j = 1, \ \text{where } w_j > 0, \ \forall j
\end{aligned} \tag{8.29}
$$

Rezaei (2016) declared the above model is an intersection of $4n - 5$ linear constraints, $2(2n - 3)$ comparison constraints, and one constraint for the weight sum of n criteria. To mimic the consistency check for pair-comparison in AHP, he also suggested a formula for calculating consistency ratio (CR) about the maximum value of the deviation ξ:

$$
\boldsymbol{CR} = \frac{\xi^*}{\text{consistency index}} \tag{8.30}
$$

For the values of the consistency index (CI), one can refer to Table 8.10, where $a_{BW} = a_{Bj} \times a_{jW}$ for perfect consistency.

In general, the values of CR are within the interval [0, 1]. When the value of ξ^* gets small, the value of CR decreases. If CR's value is less than 0.1, the pair-wise comparisons are consistent, and thus the results are reliable.

Seeing the provided steps, we know that there is nothing related to the alternative selection. Mi et al. (2019) seem to misunderstand the essence of MCDM and could

Table 8.10 Maximum value of ξ corresponding to a_{BW} values

a_{BW}	1	2	3	4	5	6	7	8	9
Consistency index (max ξ)	0.00	0.44	1.00	1.63	2.30	3.00	3.73	4.47	5.23

not supply the right classification. The applications of the stand-along BWM are only for criteria ranking and selection. If the case requires ranking and selecting alternatives, one or more MCDM approaches could be integrated with the BWM, as their classification is the multiple integration of the BWM.

We demonstrate the BWM procedure through an example.

Example 8.4 A strategic alliance partner selection problem (Chang 2021).

A leading Taiwanese green biopharmaceutical company was seeking to join multiple strategic alliances. The decision to establish an alliance was of strategic significance because the success of the alliance will have a major positive impact on the company's competitiveness. The main strategic alliance partner assessment comprised five dimensions: economic resources (X_1), innovation capability (X_2), organizational management (X_3), risk factors (X_4), and environmental protection (X_5). Each dimension comprised multiple criteria, and a total of 25 assessment criteria were included in the assessment (Chang 2021).

To clearly illustrate the BWM procedure, we only consider these five dimensions with only one expert involved. The expert was asked to choose the best criterion and the worst criterion, X_2 and X_4, respectively (Steps 1 and 2). In Step 3, the expert compared the preference of the best criterion over all the other criteria, and the resulting best-to-other (BO) vector, with numbers 1–9, is exhibited in Table 8.11. In Step 4, the expert compared the preference of all the criteria over the worst criterion, and acquired the resulting other-to-worst (OW) vector shown in Table 8.12. In Step 5, an optimal weight set is procured through a program of Eq. (8.29). The resulting optimal weight set $W^* = (0.1532, 0.4595, 0.2298, 0.0656, 0.0919)$. Then, we make a

Table 8.11 The expert's BO vector

Best	X_1	X_2	X_3	X_4	X_5
X_2	3	1	2	7	5

Table 8.12 The expert's OW vector

Worst	X_4
X_1	3
X_2	7
X_3	5
X_4	1
X_5	2

consistency check, where $\xi^L = 0$ and $\mathrm{CR} = 0$, which means the survey questionnaire is perfect consistency.

BWM has been extended in a number of different ways. Much work has been done with other MCDM techniques, including ANP (Luo et al. 2020), MOORA (Tian et al. 2018), VIKOR method (Gupta 2018) and TOPSIS (Deng et al. 2020). Furthermore, Dong et al. (2021) developed a fuzzy procedure for BWM, and other BMW studies in a group decision-making environment, such as Chang (2021), are also involved. Due to its newly developed nature, other extensions can be expected.

8.4 Summary and Discussions

TOPSIS is a popular method in MCDM development. We have addressed three related methods, i.e., fuzzy TOPSIS, VIKOR, and BWM. Fuzzy TOPSIS was introduced over four decades ago and has spawned a plethora of related works. VIKOR and fuzzy VIKOR have also inspired many papers in the last two decades. BWM is fairly recent but has been applied in quite a few studies. However, BWM could obtain the weights of criteria and could solve a MCDM problem only in combination with other MCDM techniques. The use of fuzzy sets has been expanded to different fuzzy domains, such as Pythagorean fuzzy sets, these new preference structures have been integrated into group-decision environments (Herrera-Viedma et al. 2002). As a result, modern applications are more sophisticated yet limited in their use. We must negotiate a trade-off between the mathematical beauty of the illustrations and prospective real-world applications before using such expanded use of fuzzy sets.

References

Ahuja, R.K.: Minimax linear programming problem. Oper. Res. Lett. **4**, 131–134 (1985). https://doi.org/10.1016/0167-6377(85)90017-3

Buckley, J.J., Eslami, E.: An Introduction to Fuzzy Logic and Fuzzy Sets. Springer, Heidelberg (2002). ISBN: 9783790814477

Ceballos, B., Pelta, D.A., Lamata, M.T.: Rank reversal and the VIKOR method: an empirical evaluation. Int. J. Inf. Technol. Decis. Mak. **17**, 513–525 (2018). https://doi.org/10.1142/S0219622017500237

Chang, M.H.: Using a novel hybrid MCDM model to evaluate and select the strategic alliance partners for green biopharmaceutical industry. Ph.D. Dissertation, College of Management, National Taipei University of Technology, Taipei, Taiwan (2021)

Chen, S.J., Hwang, C.L.: Fuzzy Multiple Attribute Decision Making: Methods and Applications. Springer, Berlin (1992). ISBN: 9783540549987

Chen, Y.W., Tzeng, G.H.: Using fuzzy integral for evaluating subjectively perceived travel costs in a traffic assignment model. Eur. J. Oper. Res. **130**, 653–664 (2001). https://doi.org/10.1016/S0377-2217(99)00439-7

Choquet, G.: Theory of capacities. Annales De L'institut Fourier **5**, 131–295 (1953). https://doi.org/10.5802/aif.53

Deng, F., Li, Y., Lin, H., Miao, J., Liang, X.: A BWM-TOPSIS hazardous waste inventory safety risk evaluation. Int. J. Environ. Res. Public Health **17**(5765), 1–18 (2020). https://doi.org/10.3390/ijerph17165765

Dong, J., Wan, S., Chen, S.M.: Fuzzy best-worst method based on triangular fuzzy numbers for multi-criteria decision-making. Inf. Sci. **547**, 1080–1104 (2021). https://doi.org/10.1016/j.ins.2020.09.014

Dubois, D., Perny, P.: A review of fuzzy sets in decision sciences: Achievements, limitations and perspectives. Chapter 16, 637–691. In: Greco, S., Ehrgott, M., Figueira, J.R. (Eds.) Multiple Criteria Decision Analysis: State of the Art Surveys. Springer, New York (2016). ISBN: 9781493930937

Dubois, D., Prade, H.: Fuzzy Sets and Systems: Theory and Applications. Academic Press, New York (1980). ISBN: 9780122227509

Grabisch, M.: The application of fuzzy integrals in multicriteria decision making. Eur. J. Oper. Res. **89**, 445–456 (1996). https://doi.org/10.1016/0377-2217(95)00176-X

Gul, M., Celik, E., Aydin, N., Gumus, A.T., Guneri, A.F.: A state of the art literature review of VIKOR and its fuzzy extensions on applications. Appl. Soft Comput. **46**, 60–89 (2016). https://doi.org/10.1016/j.asoc.2016.04.040

Gupta, H.: Evaluating service quality of airline industry using hybrid best worst method and VIKOR. J. Air Transp. Manage. **68**, 35–47 (2018). https://doi.org/10.1016/j.jairtraman.2017.06.001

Herrera-Viedma, E., Herrera, F., Chiclana, F.: A consensus model for multiperson decision making with different preference structures. IEEE Trans. Syst. Man Cybernet. Part A Syst. Hum. **32**, 394–402 (2002). https://doi.org/10.1109/TSMCA.2002.802821

Kaufmann, A., Gupta, M.M.: Introduction to Fuzzy Arithmetic: Theory and Applications. Van Nostrand Reinhold, New York (1991). ISBN: 9781850328810

Kaya, T., Kahraman, C.: Multicriteria decision making in energy planning using a modified fuzzy TOPSIS methodology. Expert Syst. Appl. **38**, 6577–6585 (2011). https://doi.org/10.1016/j.eswa.2010.11.081

Krishnan, A.R., Kasim, M.M., Bakar, E.M.N.E.A.: A short survey on the usage of Choquet integral and its associated fuzzy measure in multiple attribute analysis. Proc. Comput. Sci. **59**, 427–434 (2015). https://doi.org/10.1016/j.procs.2015.07.560

Lee, E.S., Li, R.-J.: Comparison of fuzzy numbers based on the probability measure of fuzzy events. Comput. Math. with Appl. **15**, 887–896 (1988). https://doi.org/10.1016/0898-1221(88)90124-1

Luo, C., Ju, Y., Santibanez Gonzalez, E.D.R., Dong, P., Wang, A.: The waste-to-energy incineration plant site selection based on hesitant fuzzy linguistic best-worst method ANP and double parameters TOPSIS approach: a case study in China. Energy **211**(118564), 1–15 (2020). https://doi.org/10.1016/j.energy.2020.118564

Mardani, A., Zavadskas, E.K., Govindan, K., Senin, A.A., Jusoh, A.: VIKOR technique: a systematic review of the state of the art literature on methodologies and applications. Sustainability **8**(37), 1–38 (2016). https://doi.org/10.3390/su8010037

Mi, X., Tang, M., Liao, H., Shen, W., Lev, B.: The state-of-the-art survey on integrations and applications of the best-worst method in decision making: why, what, what for, and what's next. Omega **87**, 205–225 (2019). https://doi.org/10.1016/j.omega.2019.01.009

Negi, D.S.: Fuzzy analysis and optimization. Ph.D. Dissertation, Department of Industrial Engineering, Kansas State University, Manhattan, Kansas (1989)

Nguyen, T.L.: Methods in ranking fuzzy numbers: a unified index and comparative reviews. Complexity, 2017: Article ID 3083745, 1-13 (1989). https://doi.org/10.1155/2017/3083745

Opricovic, S.: Multicriteria Optimization in Civil Engineering Systems. Faculty of Civil Engineering, Belgrade, 302 p. (1998). ISBN: 8680049824

Opricovic, S., Tzeng, G.-H.: Compromise solution by MCDM methods: a comparative analysis of VIKOR and TOPSIS. Eur. J. Oper. Res. **156**, 445–455 (2004). https://doi.org/10.1016/S0377-2217(03)00020-1

Palczewski, K., Sałabun, W.: The fuzzy TOPSIS applications in the last decade. Proc. Comput. Sci. **159**, 2294–2303 (2019). https://doi.org/10.1016/j.procs.2019.09.404

Ploskas, N., Papathanasiou, J.: A decision support system for multiple criteria alternative ranking using TOPSIS and VIKOR in fuzzy and nonfuzzy environments. Fuzzy Sets Syst. **377**, 1–30 (2019). https://doi.org/10.1016/j.fss.2019.01.012

Rezaei, J.: Best-worst multi-criteria decision-making method. Omega **53**, 49–57 (2015). https://doi.org/10.1016/j.omega.2014.11.009

Rezaei, J.: Best-worst multi-criteria decision-making method: some properties and a linear model. Omega **64**, 126–130 (2016). https://doi.org/10.1016/j.omega.2015.12.001

Saaty, T. L.: The Analytic Hierarchy Process. McGraw-Hill, New York (1980). ISBN: 9780070543713

Shih, H.-S.: An incremental analysis for MCDM with an application to group TOPSIS. Eur. J. Oper. Res. **186**, 720–734 (2008). https://doi.org/10.1016/j.ejor.2007.02.012

Sugeno, M.: Theory of Fuzzy Integrals and Its Applications. Ph.D. Thesis, Tokyo Institute of Technology, Tokyo (1974)

Taha, H.A.: Operations Research: An Introduction, 10th Edn. Harlow, Essex: Pearson (2017). ISBN: 9781292165547

Tan, C.: A multi-criteria interval-valued intuitionistic fuzzy group decision making with Choquet integral-based TOPSIS. Expert Syst. Appl. **38**, 3023–3033 (2011). https://doi.org/10.1016/j.eswa.2010.08.092

Tian, Z.P., Wang, J.Q., Wang, J., Zhang, H.Y.: A multiphase QFD-based hybrid fuzzy MCDM approach for performance evaluation: a case of smart bike-sharing programs in Changsha. J. Clean. Prod. **171**, 1068–1083 (2018). https://doi.org/10.1016/j.jclepro.2017.10.098

Tsaur, R.C.: Decision risk analysis for an interval TOPSIS method. Appl. Math. Comput. **218**, 4295–4304 (2011). https://doi.org/10.1016/j.amc.2011.10.001

Tzeng, G.H., Lin, C.W., Opricovic, S.: Multi-criteria analysis of alternative-fuel buses for public transportation.(2005a). https://doi.org/10.1016/j.enpol.2003.12.014

Tzeng, G.H., Ou Yang, Y.P., Lin, C.T., Chen, C.B.: Hierarchical MADM with fuzzy integral for evaluating enterprise intranet web sites. Inf. Sci. **169**, 409–426 (2005b). https://doi.org/10.1016/j.ins.2004.07.001

Yang, J.L., Chiu, H.N., Tzeng, G.H., Yeh, R.H.: Vendor selection by integrated fuzzy MCDM techniques with independent and interdependent relationships. Inf. Sci. **178**, 4166–4183 (2008). https://doi.org/10.1016/j.ins.2008.06.003

Yazdani, M., Graeml, F.R.: VIKOR and its applications: a state-of-the-art survey. Int. J. Strat. Dec. Sci. **5**, 56–83 (2014). https://doi.org/10.4018/ijsds.2014040105

Zadeh, L.A.: Fuzzy sets. Inf. Control **8**, 338–353 (1965). https://doi.org/10.1016/S0019-9958(65)90241-X

Zhang, X., Xu, Z.: Extension of TOPSIS to multiple criteria decision making with Pythagorean fuzzy sets. Int. J. Intell. Syst. **29**, 1061–1078 (2014). https://doi.org/10.1002/int.21676

Zimmermann, H.J.: Fuzzy Sets, Decision Making, and Expert Systems. Kluwer, Boston (1987). ISBN: 9780898381498

Zimmermann, H.J.: Fuzzy Set Theory—And Its Applications, 4th edn. Springer, New York (2001). ISBN: 9780898381498

Chapter 9
TOPSIS Applications

TOPSIS is one of the more popular multiple criteria modeling method in academia. Salih et al. (2019) reported detailed statistics on the number of TOPSIS journal articles published. We have roughly done an update looking at scholarly publications on EBSCO, for all time periods in their system as well as 2018–2021. These are shown in Table 9.1.

Table 9.1 demonstrates a very heavy rate of publication applying TOPSIS, to include recent years. Most of these papers were evidently on technical matters, but there are a great many published applications.

9.1 Selection Applications

This chapter will review a few of those recent papers, focusing on the weighted normalized matrices to show the criteria and relative emphasis reported. Each of these reviews will begin with a short description of the problem domain and application. They are organized to demonstrate typical applications: selection problems, with emphasis on location, assessment applications, and group applications.

One of the most popular applications of TOPSIS is to support location decisions. We review two location decisions modeled with TOPSIS. The second part will discuss selection decisions.

9.1.1 Location Decisions

In the energy arena, Otay et al. (2020) dealt with a problem involving location of a wind farm. Energy consumption continues to grow worldwide. As with many countries, Turkeyhas experienced demand growth for electricity at the rate of almost

H.-S. Shih and D. Olson, *TOPSIS and its Extensions: A Distance-Based MCDM Approach*, Studies in Systems, Decision and Control 447,
https://doi.org/10.1007/978-3-031-09577-1_9

Table 9.1 TOPSIS papers by application area

Area	Total	2018 on	Applications total	Applications 2018 on
Total	6839	3607	2324	1110
Business	1827	874	616	251
Marketing	179	91	44	21
Supply Chain Management	424	118	147	49
Logistics	431	233	156	83
Manufacturing	1386	743	417	194
Information Systems	312	159	122	39
Human Resource Management	90	31	15	4
Transportation	526	278	109	14
Safety	416	254	82	16
Health	391	236	68	15
Energy	1026	666	184	46
Water Resources	208	155	33	18

5 percent per year, and has had to import nearly 70 percent. Wind power is attractive in many places in Turkey, but investment has lagged. In recent years both public and private sectors have seen increased interest in wind power investment. Wind is attractive because:

- There is adequate wind speed in many places in Turkey, both onshore and offshore;
- Turbines don't use fossil fuels;
- Rural economies would be vitalized where wind farms were initiated, with both short-term and long-term employment opportunities;
- Generation costs would be stable;
- Locations don't affect surrounding agriculture and livestock, apart from birds.

Otay et al. (2020) evaluated four potential wind farm sites, near Balikesir, Izmir, Canakkale, and Kirklareli. They applied AHP with three experts drawn from the sponsoring power company. Criteria are given in Table 9.2:

AHP was applied along with fuzzy TOPSIS. The fuzzy aspects involved uncertainties and lack of information relative to human expert capacities. TOPSIS analysis yielded rankings recommending Balikesir, Canakkale, Izmir, and Kirklareli in that order.

Durak et al. (2021) combined AHP and TOPSIS in the analysis of technology park selection in Turkey. Technoparks promote innovation and technology development, fostering cooperation between companies and universities. The application was in Istanbul, relying upon input data from managers from companies using technoparks. The technopark was designed to increase innovation culture between technological R&D companies and entrepreneurs, aiming to support idea commercialization. The

Table 9.2 Otay et al.'s criteria for wind farms (2020)

Top level criteria	Sub-criteria
Geographical and Physical	Geomorphologic structure Topographic structure Tectonic & Seismic structure Hydrologic & Hidrogeologic structure
Natural Resources	Wind speed Wind power Effective wind hours
Economic Factors	Land cost Construction cost Investment cost ROI Operation & maintenance cost
Environmental & Social Factors	Fauna-Flora Environmental consciousness Local support Job availability Contribution to society
Legal Factors	Energy policies and regulations Licenses and trade policies
Location Factors	Distance to railways & roads Distance to electrical grid Distance to historical sites
Strategic Factors	Competitors/competition Expansion opportunities Market size

ITU ARI Technocity organization had previously signed on over 2,500 successful R&D projects resulting in 148 patents.

The analysis reported was to aid in identification of the optimal location for a new park. Decision makers were 18 authorities of the three information technology companies selected from each technopark region. AHP was used to consider fields of activity, financial status, logistic opportunities, and technological infrastructure in the study of Istanbultechnopark regions.

Durak et al. (2021) reviewed the literature to identify potential criteria. Table 9.3 displays criteria as well as the six alternative technoparks in the Istanbulregion. The 18 authorities from the IT companies evaluated criteria over these alternatives using AHP.

Input data for each technopark on each criterion are given in Table 9.4. Rent was considered a measure where more was better.

ITU had the most patents, labs, and support facilities. Bosporushad the highest score relative to cooperation and collaboration. ITU, Bosporus, Yildiz and Istanbul Technopark had the highest proximity to transport ratings. Marmara had the lopwest rent per square meter. Weightings from AHP analysis are given for criteria, as well as

Table 9.3 Technopark criteria and alternatives (Durak et al. 2021)

	Criteria	Measure		Alternatives
K1	Technopark suitability to sector	Patents	A1	ITU Technopark
K2	Infrastructure facilities	Laboratories	A2	BosporusTpark
K3	Cooperation & collaboration	Different sectors	A3	Technopark Istanbul
K4	Competitive advantage service	Support facilities	A4	Yildiz Technopark
K5	Proximity to transport	See, land, air	A5	Marmara Tpark
K6	Cost	Rent/square meter	A6	IstanbulTechnopark

Table 9.4 Input data

		K1	K2	K3	K4	K5	K6
		Patents	Labs	Sectors	Support	Transport	Rent
A1	ITU	148	445	12	25	75	62
A2	Bosporus	55	160	16	15	75	74
A3	T Istanbul	113	325	13	20	50	45
A4	Yildiz	95	286	18	18	75	52
A5	Marmara	40	125	5	14	50	30
A6	Istanbul T	35	83	4	11	75	41

the aggregate scores for each technopark site. TOPSIS was then applied. Normalized scores displayed in Table 9.5.

Criterion weights from AHP were multiplied by these scores to wield a weighted normalized matrix, which enabled identification of positive and negative ideal solution values (Table 9.6).

In Table 9.6, weights are shown in bold, the criterion weights obtained from AHP. The Positive ideals are shown in bold, the maximum weighted normalized score in each column. The Negative ideals are also in bold, the minimum weighted normalized

Table 9.5 Normalized TOPSIS matrix

		K1	K2	K3	K4	K5	K6
		Patents	Labs	Sectors	Support	Transport	Rent
A1	ITU	0.656	0.676	0.393	0.575	0.452	0.481
A2	Bosporus	0.244	0.243	0.524	0.345	0.452	0.574
A3	T Istanbul	0.501	0.494	0.425	0.460	0.302	0.349
A4	Yildiz	0.421	0.434	0.589	0.414	0.452	0.403
A5	Marmara	0.177	0.190	0.164	0.322	0.302	0.233
A6	Istanbul T	0.155	0.126	0.131	0.253	0.452	0.318

Table 9.6 Weighted normalized matrix

		K1	K2	K3	K4	K5	K6
		Patents	Labs	Sectors	Support	Transport	Rent
	Weights	**0.039**	**0.138**	**0.097**	**0.189**	**0.215**	**0.321**
A1	ITU	0.026	0.093	0.038	0.109	0.097	0.154
A2	Bosporus	0.010	0.034	0.051	0.065	0.097	0.184
A3	T Istanbul	0.020	0.068	0.042	0.087	0.065	0.112
A4	Yildiz	0.016	0.060	0.058	0.078	0.097	0.129
A5	Marmara	0.007	0.026	0.016	0.061	0.065	0.075
A6	Istanbul T	0.006	0.017	0.013	0.048	0.097	0.102
	Positive ideal	**0.026**	**0.093**	**0.058**	**0.109**	**0.097**	**0.075**
	Negative ideal	**0.006**	**0.017**	**0.013**	**0.048**	**0.065**	**0.184**

Table 9.7 Proximities to ideal solution and resulting rankings

		Ideal Dist	Nadir Dist	Proximity	Rank	AHP rank
A1	ITU	0.082	0.112	0.577	2	1
A2	Bosporus	0.133	0.056	0.294	6	4
A3	T Istanbul	0.062	0.102	0.621	1	2
A4	Yildiz	0.072	0.094	0.568	3	3
A5	Marmara	0.100	0.111	0.527	4	5
A6	Istanbul T	0.112	0.088	0.440	5	6

score for each column. Note that since cost is to be minimized, the positive ideal is the minimum and the negative ideal the maximum of the column. Table 9.7 gives the proximity values for this dataset, along with resultant rankings, compared with the rankings obtained from AHP.

AHP recommended ITU, while TOPSIS recommended Technopark Istanbul. The proximity scores indicate close results for the first four ranked alternatives. Looking at both TOPSIS and AHP, further analysis among ITU, Technopark Istanbul, and Yildiz might be useful.

9.1.2 *Selection Decisions*

Wei et al. (2019) published a report of an information systems project selection model. The popularity of the Internet has resulted in the ability to outsource information systems services, and a major industry to provide such service has evolved. Keen competition makes it possible for firms to focus their internal information systems operations on core competencies and outsource the rest. It is often the cae that development and maintenance information systems projects are outsourced, to include enterprise resource planning systems.

Wei et al. (2019) described a decision by a television assembly company in Taiwan, which needed many information systems projects to support research, production, mand testing. Management wanted to identify the most suitable projects to outsource to an Internet service provider (ISP). The company wanted to maximize total expected profit, maximize profit for the ISP, minimize development risk, and minimize administrative cost. The decision involved some interdependencies in that certain projects shared administrative, hardware, and software costs, so a linear programming model was implemented. Since multiple criteria were present, goal programming was used. TOPSIS parameters PIS and NISwere used within the goal programming model. Wei et al. looked at sensitivity to weights. Table 9.8 shows some of the tradeoffs they identified.

Table 9.8 Project selection with respect to weight adjustment (Wei et al. 2019)

Weight set	metric	E{profit}	ISP profit	Dev Risk	Admin cost	Projects selected
0.25 0.25 0.25 0.25	$p = 1$	45	1530	30	750	1 2 3 5 12 13
	$p = 2$	33	1160	23	600	1 5 12 13
	$p = 3$	33	1160	23	600	1 5 12 13
	$p = 100$	51	1700	38	930	1 2 4 5 9 12 13 14
0.4 0.2 0.2 0.2	$p = 1$	55	1950	44	970	1 2 3 5 7 8 10 12 13 14
	$p = 2$	56	2000	49	1090	1 2 4 5 6 7 8 10 12 13 14
	$p = 3$	52	1780	41	940	1 2 5 6 7 9 10 12 13 14
	$p = 100$	54	1930	47	1100	1 2 5 7 8 9 10 12 14 15
0.2 0.4 0.2 0.2	$p = 1$	55	1950	44	970	1 2 3 5 7 8 10 12 13 14
	$p = 2$	56	2010	49	1100	1 2 4 5 7 8 10 12 13 14
	$p = 3$	53	1920	45	1090	1 2 4 5 7 8 9 10 12 14
	$p = 100$	51	1700	38	930	1 2 4 5 9 12 13 14
0.2 0.2 0.4 0.2	$p = 1$	32	1220	22	630	1 2 3 5
	$p = 2$	27	1050	19	600	1 5 12
	$p = 3$	27	1050	19	600	1 5 12
	$p = 100$	27	1050	19	600	1 5 12
0.2 0.2 0.2 0.4	$p = 1$	45	1530	30	750	1 2 3 5 12 13
	$p = 2$	24	970	17	530	1 2 5
	$p = 3$	26	1060	20	590	1 2 5 14

The p values in Table 9.8 are the power metrics applied within TOPSIS distance functions. $p = 1$ is linear, $p = 2$ Euclidean, $p = 3$ third power, and $p = 100$ used to represent minmax (metric ∞). The results in terms of projects selected are variable, due to the integer goal programming model output. Projects 1 and 5 were selected in all models displayed, while project 12 appeared in most weight sets. Emphasizing minimization of development risk (weight set 0.2 0.2 0.4 0.2) and minimizing administrative costs (weight set 0.2 0.2 0.2 0.4) had a clear impact in reducing the number of projects selected.

Rehman and Ali (2021) applied fuzzy TOPSIS to identify crude oil import transportation route options on the Belt and Road initiative. One driver of that program is reduction of dependence on fossil fuel imports, which have increased with the massive increase in Chinese economic activity. Railways and pipelines are being used as modes of transportation for petroleum, along with sea transport through the Strait of Malacca, which is a choke point with shipping risks. The China Pakistan Economic Corridor (CPEC) is a flagship project aimed at connecting Western China with the Arabian Sea. Deep sea ports in Pakistanat Gwadar and Karachiare key elements of this program. Infrastructure including power stations, highway networks, and upgraded rail networks in Pakistanlink Gwadar and Karachito Kashghar city in Western China.

Rehman and Ali's research (2021) aimed to identify the optimal combination of highways, railways, and pipelines and to compare the viability of the maritime route passing through CPEC with currently operating systems. Alternatives were the various combinations of transport mechanisms, while criteria included security risks as well as benefits and costs. Because of the uncertainty of inputs, fuzzy TOPSIS was applied. The first phase of the analysis applied TOPSIS to compare highways, railways, and pipelines to get to Qinghaifrom either Karachior Gwadar (Table 9.9).

Table 9.9 Belt initiative alternatives to get to seaports (Rehman and Ali 2021)

	Criteria	Units	Highways	Rail	Pipeline
Karachi	Time	Days	2.22	1.81	1.44
	Costs	US$	4.54	6.81	8.93
	Energy	Tons	0.15	0.23	0.10
	Emissions	CO_2/bbl	970,272	573,704	658,449
	Risks	%	19	18	19
	TOPSIS	Rank	2	3	1
Gwadar	Time	Days	2.81	2.47	2.10
	Costs	US$	5.75	8.16	10.28
	Energy	Tons	0.19	0.28	0.15
	Emissions	CO_2/bbl	1,227,042	861,378	946,123
	Risks	%	52	57	58
	TOPSIS	Rank	2	3	1

Table 9.10 Maritime option data

Africa	Criteria	Units	Shanghai	Myanmar	Karachi	Gwadar
	Time	Days	29.66	22.39	31.3	20.8
	Costs	US$	10.76	8.74	12.7	11.3
	Energy	Tons	519,089	414,847	342,832	345,636
	Emissions	CO_2/bbl	1,106,918	306,462	965,835	678,324
	Risks	%	77	73	52	52
	TOPSIS	Rank	4	1	3	2
MidEast	Time	Days	18.35	10.22	3.7	3.9
	Costs	US$	9.3	7.2	10.5	9.2
	Energy	Tons	325,317	181,601	33,548	46,776
	Emissions	CO_2/bbl	10,975,775	293,050	948,052	661,139
	Risks	%	33	30	9	9
	TOPSIS	Rank	4	3	2	1
America	Time	Days	35.65	32.2	29.77	30.5
	Costs	US$	11.62	10	13.6	12.48
	Energy	Tons	665,026	589,888	479,985	509,512
	Emissions	CO_2/bbl	1,115,309	316,526	1,254,641	603,001
	Risks	%	91	87	67	67
	TOPSIS	Rank	4	1	3	2

The next phase was to analyze maritime routes passing through Gwadar, Karachi, Myanmar, and Shanghai from sources in Africa, the Middle East, or America (data in Table 9.10).

Sensitivity ranked Myanmaras the best choice, followed by the CPEC routes. The Eastern China route (Shanghai) was ranked last in all cases. Incorporating the CPEC route in China's oil import framework was expected to increase bypass options and reduce dependence on the Straits of Malacca, and would also upgrade economic development of Xinjiang and western Chinese providences.

9.1.3 Other Applications

Mohammed (2020) applied fuzzy TOPSIS to the supplier selection problem, a key issue in supply chain management. Traditional criteria included cost, healthy meat, technology capability, and delivery capability. Green criteria considered were environmental impact, waste treatment capability, and pollution impact. Social criteria involved safety, staff development, and information disclosure. The model could be applied to whatever suppliers were available, on a case-by-case basis.

Patil et al. (2020) also applied fuzzy TOPSIS to the supplier selection in the fiber industry located in Goa, India. They considered both qualitative (quality, reliability, flexibility, stability, capability and availability) and quantitative (price, order quantity, and credit/delivery/time) criteria. Four decision makers provided assessment of qualitative criteria. The analysis covered three suppliers as the alternatives.

Chen et al. (2021) used data envelopment analysis along with TOPSIS to consider quantity discounts in automobile component procurement. Thirty suppliers were screened for capabilities and performance, considering eleven criteria grouped into the four components of quality, service, technology and cost.

9.2 Assessment Applications

TOPSIS has also been widely applied to assessment of systems. Vasilić et al. (2020) reported an application in Information/Computer Technology Selection. Entropy was used in conjunction with TOPSIS in evaluating European country use of information and computer technology (e-commerce, internet connection, website/social media, and e-business). Sixteen subcriteria were used to cluster 31 European countries. This study evaluated information and computer technology (ICT) systems in 31 European countries. TOPSIS was used to develop an enterprises index based on entropy. Table 9.11 gives the criteria use.

Table 9.11 ICT usage criteria (Vasilić et al. 2020)

Criteria	Description	Type	Global weight
E-Commerce	Sales to other EU countries & the world	Max	0.07293
	E-commerce sales	Max	0.04365
	Total turnover	Max	0.08796
	Costs/complaints, issues	Min	0.10639
Internet Connection	Internet access	Max	0.00062
	Employees with www access	Max	0.02375
	Maximum download speed	Max	0.01636
	Employees with mobile devices	Max	0.05756
Websites/Social Media	Enterprises with a web	Max	0.01054
	Enterpriseblogs	Max	0.13578
	Image and marketing	Max	0.02940
	Customer involvement	Max	0.04828
E-Business	Software solutions	Max	0.04804
	Automated billing process	Max	0.14410
	Cloud services	Max	0.10480
	3D printing	Max	0.06982

A group decision-making platform was used to aggregate fuzzy input for step 3 of measuring performance. Group decision makers using Likert categories. The aggregation of group member ratings was not presented. Each decision maker was given a weight representing relative importance among the group. Decision makers evaluated both criterion weights as well as alternative scores on each criterion using interval-valued fuzzy numbers. Then TOPSIS was applied.

Song et al. (2020) demonstrated evaluation of investment risk using criteria of investment amount, income, cost, tax expenditure and investment duration. Song et al. gave a hypothetical TOPSIS model to compare investment projects. The criteria considered were investment, cost, taxation, income and duration. Input data was fuzzy. Açikgöz (2021) evaluated cash flow profiles of Turkish manufacturing enterprises with TOPSIS using ten traditional financial ratios. Firms were evaluated at five lice cycle stages. Ouyang et al. (2014) applied entropy within TOPSIS to measure quality characteristics in assessing manufacturing quality with the intent of evaluating firm ability to meet six sigma standards. Three experts provided ratings over three quality characteristics. TOPSIS was compared with the process capability analysis chart method, and with the author proposed quality characteristic analysis chart.

In the service area, Cui et al. (2020) modeled public services provided to rural areas in the agricultural domain using TOPSIS. Thirteen Chinese regions were assessed using eleven first level criteria (39 s level measures). Hejazinia (2021) applied fuzzy TOPSIS to evaluation of electronic service risk in information and communication technology. The high level risks assessed were technological (with two submeasures), operational (three sub-measures), environmental (four submeasures), and organizational (three sub-measures). Risk analysis was also modeled by Li et al. (2020) who applied failure mode and effects analysis considering occurrence, severity and detection to assess failure risks. The inputs were provided by a designer, an assembler, a quality control engineer, and an operator. Stoyanets et al. (2020) evaluated rural governance capacity in Chinausing the entropy weight method within TOPSIS. Rural governance capacity was assessed for 300 villages on public service capacity (16 measures), resource guarantee capacity (9 measures), villager participation capacity (5 measures) and business participation (3 measures).

There also have been many TOPSIS applications in assessing manufacturing systems. Ban et al. (2020) used fuzzy AHP and TOPSIS in evaluating 33 Romanian manufacturing companies, considering eight financial and seven non-financial criteria. Fuzzy TOPSIS was also applied by Liu et al. (2017) to assess lean transformation of firms. The study used 12 experts to provide inputs in the comparison of two companies.

9.3 Group Applications

Foreign direct investment is considered a key element to Turkish economic growth. Çalik et al. (2019) developed a group AHP/TOPSIS model to aid in selecting investment opportunities. They gave an example with 3 criteria with a total of nine sub-criteria with five expert decision makers to select the most suitable sector in the Turkish economy for investment. The experts were obtained from academia as well as the financial and investment industries.

The criteria consisted of economic, political and country factors. AHP was used to generate weights using a fuzzy triangular scale. Alternatives were scored on each of the nine sub-criteria shown in Table 9.12.

The Çalik et al. analysis followed TOPSIS with sensitivity analysis by varying weights.

Jiang et al. (2019) proposed a decision-making approach to extend fuzzy-based TOPSIS model. They applied this approach in a group decision-support context to the problem of selecting suppliers. Their framework involved 8 steps:

1. Problem definition: To select the optimal supplier with the highest resilience from among a group of available suppliers;
2. Measuring performance: Identify evaluation criteria relative to supplier resilience;
3. Measuring performance: Gather data
4. Scoring alternatives: Evaluate supplier performance over each criterion;
5. Weighting criteria and decision makers
6. Calculating aggregate scores: To include total value to rank from TOPSIS
7. Dealing with uncertainty: Perform sensitivity analysis
8. Report and Monitor

Evaluation criteria are given in Table 9.13.

Table 9.12 Çalik et al. (2019) model data

Criteria	Subcriteria	Alternatives
Economic	High inflation	Wholesale & retail
	Exchange rate	Energy
	Wage costs	Information/communication technologies
Political	Political risk	Finance/banking
	Investment openness	Manufacturing
	War/terror risk	
Country	Geography	
	Culture	
	Clustering	

Table 9.13 Jiang et al.'s evaluation criteria (2019)

Type	Criteria	Description	Type	Direction
Resiliency	Supply chain density	Quantity and geographical spacing of nodes	Linguistic	Min
	Supply chain complexity	Number of nodes and interconnections	Linguistic	Max
	Responsiveness	Response speed to market demand	Linguistic	Max
	Number of critical nodes	Relative importance of a given node	Numeric	Min
Critical	Re-engineering	Corrective procedure to incorporate product design change	Linguistic	Max
	Buffer capacity	Extra stock to mitigate risk	Numeric	Max
	Supplier resource flexibility	Logistics strategies to release product or to procure components	Linguistic	Max
	Lead time	Delay between initiation and execution	Numeric	Min

9.3.1 *Effectiveness*

Zhuang et al. (2019) presented a case involving evaluating the effectiveness of a senior center, an issue of growing concern in a world with an aging population. AHP was applied to generate weights followed by TOPSIS. Three senior individuals facing selection of a retirement residence provided input over four alternative shared residences. Five criteria were considered (with fourteen sub-criteria—see Table 9.14).

Note that the sub-criteria Neatness and Ventilation have perfect scores for all four alternatives. However, evaluation models can be viewed as usable for consideration of other alternatives, so while these sub-criteria won't play any role in ranking these four alternatives, they might well play a role for other applications.

AHP was used to rank the five criteria by each decision maker. Each decision maker applied AHP to the five criteria, and the geometric mean used in the aggregation. Each decision maker also applied AHP to score alternative performances over each criterion.

9.4 Conclusions

We have reviewed eight specific TOPSIS applications from among thousands. Those selected are recent, selected for our view of interesting applications. Table 9.15 shows a comparative layout:

Table 9.14 Zhuang et al.'s criteria (2019)

Criteria	Sub-criteria	Measure	First	Second	Third	Fourth
Finance	Price	NT$/mo	20,900	19,000	16,000	12,800
	Support	Score	50	0	50	0
Environment	Neatness	Score	100	100	100	100
	Use space		30	16	22	23.17
	Safety	Score	100	80	80	80
	Bedroom	Score	50	75	50	50
Facilities	Ventilation	Score	100	100	100	100
	Facilities	Score	83	83	100	67
	Entertainment	Score	86	100	71	86
Servicer	Healthcare	Score	100	100	60	80
	Staff	Score	100	100	83	83
	Food	Score	75	75	100	50
Convenience	Transportation	Score	75	100	100	50
	Living	Score	100	80	100	100

Table 9.15 Chapter 9 Applications

Application	Paper	Application	Features
Location	Otay et al. (2020)	Wind power farm selection	fuzzy AHP & fuzzy TOPSIS
	Durak et al. (2021)	Technopark selection	AHP compared to TOPSIS
Selection	Wei et al. (2019)	Information project selection	TOPSIS and goal programming
	Rehman & Ali (2021)	Oil import route selection	Cost benefit analysis of alternative routes
Assessment	Vasilić et al. (2020)	ICT evaluation	ICT usage in enterprises index
Group	Çalik et al. (2019)	Foreign direct investment	AHP for weights, TOPSIS for ranking
	Jiang et al. (2020)	Supplier selection	Crisp and Interval valued fuzzy sets
	Zhuang et al. (2019)	Senior center effectiveness	Intuitionist fuzzy group multicriteria

TOPSIS has been used for a number of multiple criteria applications, many selection decisions, but also many studies have applied TOPSIS to assess performance. This chapter gave examples of both types of application, with discussion of analysis frameworks. References can be read to find greater details.

References

Açikgöz, T.: Investigation of cash flow profiles of manufacturing sector and comparison of financial performances with TOPSIS method. J. Account. Finance **90**, 127–147 (2021). https://doi.org/10.25095/mufad.837360

Ban, A.I., Ban, O.I., Bogdan, V., Sabau Popa, D.C., Tuse, D.: Performance evaluation model of Romanian manufacturing companies by fuzzy AHP and TOPSIS. Technol. Econ. Dev. Econ. **26**(4), 806–836 (2020). https://doi.org/10.3846/tede.2020.12367

Çalik, A., Çizmecioğlu, S., Akpinar, A.: An integrated AHP-TOPSIS framework for foreign direct investment in Turkey. J. Multi-Criteria Decis. Anal. **26**(5/6), 296–307 (2019). https://doi.org/10.1002/mcda.1692

Chen, J., Xu, Z., Gou, H.D., Zhang, J.: Automobile components procurement using a DEA-TOPSIS-FFMIP approach with all-unit quantity discount and fuzzy factors. Technol. Econ. Dev. Econ. **27**(2), 311–352 (2021). https://doi.org/10.3846/tede.2020.13176

Cui, L., Zhou, H., Kolivand, H., Balas, V.E., Paul, A., Ramachandran, V.: Spatial feature analysis of rural basic public service supply based on TOPSIS model and data aggregation algorithm. J. Intell. Fuzzy Syst. **39**(4), 5623–5634 (2020). https://doi.org/10.3233/JIFS-189042

Durak, I., Arslan, H.M., Őzdemir, Y.: Application of AHP-TOPSIS methods in technopark selection of technology companies: Turkish case. Technol. Anal. Strat. Manag. 1–15 (2021). https://doi.org/10.1080/09537325.2021.1925242

Hejazinia, R.: Electronic service risk pattern and ranking by using fuzzy TOPSIS technique. Int. J. Manag. Account. Econ. **8**(3), 143–155 (2021). https://doi.org/10.5281/zenodo.4698992

Jiang, D., Hasan, M.M., Faiz, T.I., Noor-E-Alam, M.: A possibility distribution-based multicriteria decision algorithm for resilient supplier selection problems. J. Multi-Criteria Decis. Anal. **27**(3/4), 203–223 (2020). https://doi.org/10.1002/mcda.1696

Li, G.-F., Li, Y., Chen, C.-J., He, J.-L., Hou, T.-W., Chen, J.-H.: Advanced FMEA method based on interval 2-tuple linguistic variables and TOPSIS. Qual. Eng. **32**(4), 653–662 (2020). https://doi.org/10.1080/08982112.2019.1677913

Liu, C.-C., Niu, Z.-W., Chang, P.-C., Zhang, B.: Assessment approach to stage of lean transformation cycle based on fuzzy nearness degree and TOPSIS. Int. J. Prod. Res. **55**(23), 7223–7235 (2017). https://doi.org/10.1080/00207543.2017.1355124

Mohammed, A.: Towards a sustainable assessment of suppliers: an integrated fuzzy TOPSIS-possibilistic multi-objective approach. Ann. Oper. Res. **293**(2), 639–668 (2020). https://doi.org/10.1007/s10479-019-03167-5

Otay, I., Jaller, M., Kahraman, C.: A novel pythagorean fuzzy AHP and TOPSIS method for the wind power farm location selection problem. J. Intell. Fuzzy Syst. **39**(5), 6193–6204 (2020). https://doi.org/10.3233/JIFS-189089

Ouyang, L.-Y., Chen, K.-S., Yang, C.-M., Hsu, C.-H.: Using a QCAC-entropy-TOPSIS approach to measure quality characteristics and rank improvement priorities for all substandard quality characteristics. Int. J. Prod. Res. **52**(10), 3110–3124 (2014). https://doi.org/10.1080/00207543.2013.865092

Patil, A.N., Shiykumar, K.M., Manjunath Oatel, G.C., Jatti, S.P., Rivankar, S.N.: Fuzzy TOPSIS and grey relation analysis integration for supplier selection in fiber industry. Int. J. Supply Oper. Manag. **7**(4), 373–383 (2020). https://doi.org/10.22034/IJSOM.2020.4.6.

Rehman, O.U., Ali, Y.: Optimality study of China's crude oil imports through China Pakistan economic corridor using fuzzy TOPSIS and cost-benefit analysis. Transp. Res. Part E **148**, 1–17 (2021). https://doi.org/10.1016/j.tre.2021.102246

Salih, M.M., Zaidan, B.B., Zaidan, A.A., Ahmed, M.A.: Survey on fuzzy TOPSIS state-of-the art between 2007 and 2017. Comput. Oper. Res. **104**, 207–227 (2019). https://doi.org/10.1016/j.cor.2018.12.019

Song, Y., Li, X., Li, Y., Hong, X.: Assessing the risk of an investment project using an improved TOPSIS method. Appl. Econ. Lett. **27**(16), 1334–1339 (2020). https://doi.org/10.1080/13504851.2019.1679339

Stoyanets, N., Hu, Z., Chen, J., Niu, L.: Managing sustainability development of the agricultural sphere based on the entropy weight TOPSIS model. Int. J. Technol. Manag. Sustain. Dev. **19**(3), 263–278 (2020). https://doi.org/10.1386/tmsd_00026_1

Vasilić, N., Semenčenko, D., Popović-Pantić, S.: Evaluating ICT usage in enterprises in Europe: TOPSIS approach. Econ. Themes **58**(4), 529–544 (2020). https://doi.org/10.2478/ethemes-2020-0030

Wei, C.C., Chang, Y.-L., Lee, K.-L.: How to select suitable manufacturing information system outsourcing projects by using TOPSIS method. Int. J. Prod. Res. **57**(13), 4333–4350 (2019). https://doi.org/10.1080/00207543.2019.1572930

Zhuang, Z.-Y., Su, C.-R., Chang, S.-C.: The effectiveness of IF-MADM (Intuitionistic-fuzzy multi-attribute decision-making) for group decisions: methods and an empirical assessment for the selection of a senior centre. Technol. Econ. Dev. Econ. **25**(2), 322–364 (2019). https://doi.org/10.3846/tede.2019.8399

Index

H.-S. Shih and D. Olson, *TOPSIS and its Extensions: A Distance-Based MCDM Approach*, Studies in Systems, Decision and Control 447,
https://doi.org/10.1007/978-3-031-09577-1

The manufacturer's authorised representative in the EU is Springer Nature Customer Service Centre GmbH, Europaplatz 3, 69115 Heidelberg, Germany. If you have any concerns regarding our products, please contact ProductSafety@springernature.com

Printed and bound by CPI Group (UK) Ltd, Croydon, CR0 4YY
30/09/2026
02225462-0003